D1083894

Videoconferencing and Videotelephony

Technology and Standards

Second Edition

For a complete listing of the *Artech House Telecommunications Library,* turn to the back of this book.

Videoconferencing and Videotelephony

Technology and Standards

Second Edition

Richard Schaphorst

Artech House
Boston • London

ADH-5497

Library of Congress Cataloging-in-Publication Data
Schaphorst, Richard.
    Videoconferencing and videotelephony : technology and standards
/ Richard Schaphorst. — 2nd ed.
        p.   cm. — (Artech House telecommunications library)
    Includes bibliographical references and index.
    ISBN 0-89006-997-2 (alk. paper)
    1. Interactive multimedia.   2. Videoconferencing.   3. Digital
communications.   I. Title. II. Series.   QA
    QA76.76.I59 S33   1999   76.76
    621.39'9—dc21   .I59   99-17753
                                       S33   CIP
                                       1999

British Library Cataloguing in Publication Data
Schaphorst, Richard
    Videoconferencing and videotelephony : technology and standards. — 2nd ed.
    1. Videoconferencing 2. Video telephone 3. Videoconferencing — Standards
    4. Video telephone — Standards
    I. Title
    621.3'85

    ISBN 0-89006-997-2

Cover design by Lynda Fishbourne

© 1999 ARTECH HOUSE, INC.
685 Canton Street
Norwood, MA 02062

International Standard Book Number: 0-89006-997-2
Library of Congress Catalog Card Number: 99-17753

10 9 8 7 6 5 4 3 2 1

Contents

Acknowledgments

This book frequently references telecommunication standards that have been, and are continuing to be, developed by the ITU and ISO standards organizations. Consequently, I wish to acknowledge the work of all the experts and organizations that have contributed to these standards over the years. I also wish to acknowledge a few particular individuals for their contributions to the book. Richard Cox was a major contributor to Chapter 4 on audio coding. Gary Thom is the primary author of Chapter 7 on H.323 and provided update material on Chapter 10 (T.120). The major update on broadband ISDN terminals, in Chapter 9, was developed by Corey Gates. Neil Randall has provided many contributions over the years, and Ronda Diggs was a key resource in the processing of text and figures. Finally, I would like to thank Barbara Lovenvirth, and other members of the staff from Artech, for their patience as this second edition unfolded. I thank you all.

1

Introduction

Our lives have been revolutionized by three developments from the world of electronic communications: the telephone, the television, and the computer. The telephone has been a pervasive fundamental communications tool in both the home and business environments for many decades. Although life at home has been turned upside down by broadcast television for many years, TV is just beginning to seriously impact the business world. The computer is the most recent explosion in our daily lives, giving rise to vast quantities of "data" to be shared. These three signals—audio, video, and data—are now being merged under the new banner of multimedia.

This book describes recent developments in the technology and standards for the communication of multimedia signals. Examples of multimedia communications include the videophone (VP), video teleconferencing (VTC) (in the conference room as well as on the desktop), remote access to multimedia databases, distance learning, telemedicine, telecommuting, and telemarketing. Although much of the technology, and many of the standards, covered in this book are applicable to the world of entertainment TV, this area is not the focus. Instead, this book is concerned with person-to-person, or person-to-database, interactive multimedia communications. Communication channels used to carry these multimedia signals include the telephone network, the integrated services digital network (ISDN), local area networks (LANs), and mobile networks.

The markets for video teleconferencing and the videophone are growing rapidly. There are three major reasons for this: improved audiovisual quality, reduced cost, and communication standards. Recent breakthroughs in video and

1

audio compression technology are responsible for the improvement in quality. An overview of video and audio compression technology is provided in Chapters 3 and 4, respectively. The cost of video teleconferencing and videophone systems has been drastically reduced in two fundamental areas: communications and the VTC/VP terminal itself. The communication cost has been radically reduced because the transmission bit rate has dropped sharply, and the cost per bit from the common carrier has also been cut. VTC/VP terminal cost has gone down because of the incredible strides in integrated circuit development. Last, but not least, the VTC/VP revolution could not have occurred without the development of communication standards. A few years ago the VTC/VP market was frozen because different vendor's terminals could not talk to each other. They each used different proprietary coding algorithms. In 1990, the International Telecommunications Union (ITU) finalized the H.320 VTC/VP standard, and today every manufacturer provides the standard in the codec. The standard accomplishes a number of objectives. It assures interoperability between terminals manufactured by different vendors. It also reduces terminal cost because chip manufacturers produce devices implementing the standard in high volume—therefore low cost.

The market for video teleconferencing and videophone systems has several dimensions. The VTC market in the business community is relatively mature and is rapidly growing. In this application the transmission bit rates range from 128 Kbps for lower-level elements of an organization up to 384 Kbps for management levels. Fixed and roll-around units are installed in conference rooms, while systems using the personal computer platform are proliferating on the desktop. Growth in the desktop area is being spurred by the recent acceptance, and recognized productivity, of telecommuting. Workers are using their personal computers at home as platforms for desktop videoconferencing to the office, and worldwide. In many of these applications, it is recognized that the data element of the multimedia communications (for example, whiteboard, spreadsheet, document editing) is extremely important, perhaps more important than motion video.

Wide area networks (WANs) have been implemented by circuit-switched technology from their inception and continue to dominate today. However, a revolution in long-distance communication, based upon packet technology, is underway. The Internet and broadband ISDN/asynchronous transfer mode (ATM) are prime examples of these new packet-based networks. The H.323 multimedia standard has been recently developed by the ITU to enable communication over these new networks as well as legacy LAN networks such as Ethernet and token ring. H.323 terminals are being introduced into the market slowly and cautiously because (1) the quality of service over the Internet is still uncertain and evolving, (2) the existing LAN network infrastructure cannot absorb a sudden influx of video traffic, and (3) H.323 typically calls for gate-keeper and gateway devices as well as the end-point terminal itself which complicates the implementation.

The market for the videophone is not as mature as video teleconferencing due to the marginal quality that has been available from recent products and the high cost. This will change rapidly because of the quality improvement inherent in H.324 and H.320 standards as well as the price reduction that will result from mass production of standardized products. The H.324 videophone has been rapidly adopted for use in the home by the consumer. Major applications include the person-to-person videophone, security, home healthcare, and remote monitoring.

The major part of this book is concerned with the new standards that define the procedures for multimedia communications. Figure 1.1 and Table 1.1 illustrate the key ITU recommendations that have been, and are being, developed for VTC/VP. The H.320 standard, designed for operation over the narrowband ISDN (N-ISDN), is the most mature standard (established in 1990), and forms the cornerstone of all videoconferencing systems—room-based as well as desktop. H.320 is available from all vendors and guarantees interoperability between systems from different manufacturers.

Recommendations H.323, H.324, H.310, and H.322 are relatively new, having been fully approved in 1996. H.324 defines a multimedia terminal that communicates speech, data, and video signals over the public switched telephone network. As part of the H.324 terminal, the new standards established for speech coding, video coding, control, and multiplex are G.723.1, H.263, H.245, and H.223, respectively.

H.321 and H.310 are new recommendations defining videoconferencing terminals for transmission via the broadband ISDN (B-ISDN)/ATM network. H.321 merely converts the H.320 standard from N-ISDN operation to B-ISDN transmission; all of the H.320 infrastructure (H.261, H.221, H.242) remains intact to maximize interoperability between the N-ISDN and B-ISDN networks. H.310 is a new recommendation that adapts the ISO MPEG-2 standards for communication over the B-ISDN/ATM network. H.262 and H.222.0 are common text standards with ISO MPEG-2.

H.322 adapts the H.320 standard for those LAN networks that guarantee the bandwidth (for example, ISO Ethernet). H.323 is the new standard designed to provide videoconferencing over nonguaranteed-bandwidth LANs such as Ethernet and token ring. The H.323 standard was fully approved in 1996.

The reader should note the H.261 video coding standard is mandatory for all the recommendations listed in Table 1.1. This greatly enhances interoperability between networks. A great deal of commonality also exists in the audio, multiplex, and control standards.

All the aforementioned recommendations, including those listed in Figure 1.1 and Table 1.1, define multimedia terminals that provide for the transmission of audio, data, and video signals. In many applications, particularly those for the desktop, data originating on the PC/workstation platform is the most important information to be transmitted. The ITU has developed the T.120 series of recommendations for the communication of this data. T.120 defines a protocol stack that can be used for data-only transmission; in addition, the T.120 protocol is used for the data component in multimedia terminals such as H.320, H.324, H.321, and H.322.

All of the above standards are interactive and require two-way communication channels for operation. The ITU has also created the H.331 standard to provide for the broadcast of multimedia signals for applications such as distance education and business television.

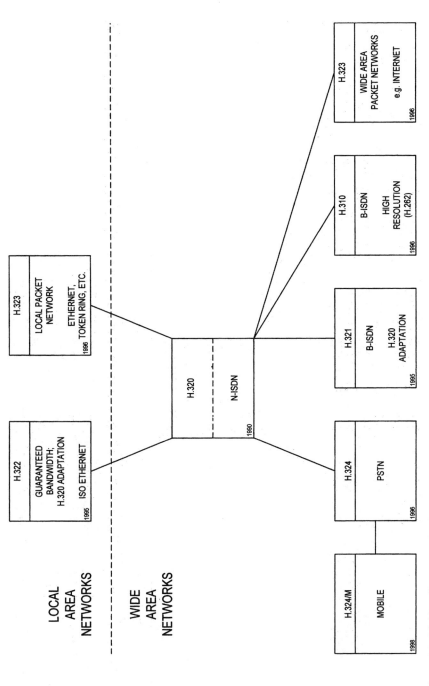

Figure 1.1 Multimedia communication standards.

Table 1.1
ITU Audiovisual Recommendations

	H.320	H.323	H.324	H.321	H.310	H.322
Approval Date	1990	1996	1996	1995	1996	1995
Network	Narrowband switched digital N-ISDN	Non-guaranteed bandwidth packet networks (Ethernet, Internet)	PSTN, mobile, N-ISDN	B-ISDN ATM	B-ISDN ATM	Guaranteed bandwidth packet switched networks (ISO Ethernet)
Video	H.261 (M) H.263	H.261 (M) H.263	H.261 (M) H.263 (M)	H.261 (M) H.263	H.261 (M) H.262*	H.261 (M) H.263
Audio	G.711 (M) G.722 G.728	G.711 (M) G.722 G.728 G.723.1 G.729	G.723.1 (M)	G.711 (M) G.722 G.728	G.711 (M) G.722 G.728 MPEG 1, 2	G.711 (M) G.722 G.728
Multiplexing	H.221	H.225.0	H.223	H.221	H.222.0/H.222.1	H.221
Control	H.230 H.242	H.245	H.245	H.242	H.245	H.242 H.230
Multipoint	H.231 H.243	H.323		H.231 H.243	H.b multipoint	H.231 H.243
Data	T.120	T.120	T.120	T.120	T.120	T.120

Table 1.1 *(continued)*

	H.320	H.323	H.324	H.321	H.310	H.322
Security	H.233 H.234	H.235		H.233 H.234		H.233 H.234
Comm. Interface	I.400	TCP/IP	V.34 modem	AAL I.363 AJM I.361 PHY I.400	AAL I.363 AJM I.361 PATS I.400	I.400 & TCP/IP

*—Common text with ISO MPEG2

M—Mandatory

2

Video Teleconferencing: Benefits and System Design

2.1 Introduction

A teleconference is a meeting between people who are physically separated from each other, achieved by using electronic communication techniques. Teleconferencing systems can be broadly classified into three categories: audio only, audiographic, and full motion video. Of course, a full motion video system includes audio and can include graphics as an option.

As a further classification, there are two types of full motion video teleconferencing systems, interactive and broadcast business television. In the case of broadcast business television, the typical application is a single event (for example, a product announcement) with one transmitter and many receivers. The primary focus of this chapter is interactive videoconferencing as opposed to one-time broadcast events. Interactive videoconferencing includes groups of people in a conference room or a single person using a desktop videophone.

2.2 Benefits of Video Teleconferencing

The benefits that organizations can gain from videoconferencing are outlined as follows.

- *Faster decision making:* Since a videoconference can be established with little more difficulty than an in-house meeting, people separated by miles can come together and share ideas and information when the need arises. There is no need to delay a decision until all participants can clear days for travel on their calendars. In industry, this is particularly valuable in rapidly resolving unexpected crises.

- *Better decisions:* Because there is no additional cost for additional meeting participants, the conference can include all the people that are involved in the undertaking. If travel were involved, this might not be possible, even if the organization were willing to incur the cost. Also, information sources not originally considered may be brought into the meeting. People may join a videoconference as easily as they can an in-house meeting.

- *Increased productivity:* Valuable employees need not waste time traveling to and from airports, waiting in check-in lines, and sitting on the tarmac awaiting takeoff clearance. If a meeting takes two hours, the participants will expend only two hours of their valuable time.

- *More meetings:* VTC permits an organization to have more meetings than otherwise would be feasible. More meetings to manage a project, for example, could result in a higher quality result or more rapid completion.

- *Increased employee safety:* Travel involves some risk. A videoconference eliminates this concern.

- *Tighter security:* If coworkers travel together and discuss business, information, even though unclassified, may be compromised if their discussions are overheard.

- *Improved employee morale:* Some travel is enjoyable. However, frequent trips, particularly to remote locations or second-tier cities, can be depressing to employees. Videoconferencing cannot eliminate the need for travel, but it can reduce travel requirements significantly.

- *Avoided travel costs:* This is the most easily quantified benefit. Many costs are associated with official trips. The traveler must pay for conveyance to the airport or pay for parking. At the destination city, there may be the need for a rental car or a taxi. Some trips require an overnight stay in a hotel. Nearly all trips call for payment of meal costs, and, of course, the cost of an airline ticket or alternate transportation must be included. Employees can seldom take advantage of discount rates that require advance payment or weekend stays.

- *Reduced fatigue:* Business travel is frequently a tiring, frustrating experience that can result in fatigue and consequential poor performance during the meeting as well as after the traveler has returned home. In the case of long distance travel, it can take several days for this fatigue, or jet lag, to diminish. Upon occasion, the employee is truly sickened by the trip. Obviously, the elimination of this fatigue is an important VTC benefit that is difficult to quantify.

- *Efficient use of key personnel:* One of the VTC benefits that is difficult to quantify is the ability to efficiently employ and apply critical personnel. For example, there are cases in industry where it would be desirable for a particular key person to attend a meeting in Europe in the morning and in Japan in the afternoon. In effect, VTC makes this feasible.

- *More disciplined, productive meetings:* In many cases, more is accomplished at a VTC meeting than at a face-to-face meeting because all participants realize that it is necessary to operate in a more disciplined manner at the VTC meeting. There are fewer interruptions, and more listening. One reason for this discipline is the fact that most VTC meetings have a fixed time duration.

- *Team building:* When meetings are held at the normal frequency, it is typically necessary for a project to be organized on a basis where all decisions and control originate at the top and are fed down throughout the organizational structure. This is often necessary to meet program schedule requirements. Since VTC permits more frequent meetings, it becomes possible to more frequently reach decisions by consensus rather than top down, thereby achieving a sense of a team effort, which in turn results in a greater degree of motivation and energy by all members of the organization.

- *Interning:* It is frequently convenient to include nonparticipatory observers in a VTC because it can be done with very little cost relative to face-to-face meetings. This provides an inexpensive way to train personnel by exposing them to a particular project and management procedures.

- *Reliability:* In many cases, a face-to-face meeting is canceled, or its effectiveness reduced, because of terminated flights or bad weather, for example. VTC eliminates this problem.

2.3 Video Teleconferencing Overview

At the highest level, a VTC system consists of a number of VTC terminals, or nodes, interconnected by means of a communication network. Such a configuration is illustrated in Figure 2.1. Full duplex communication channels are used to create a real-time interactive environment (for example, motion video, audio, and graphics) between remote sites. The VTC terminals can vary greatly

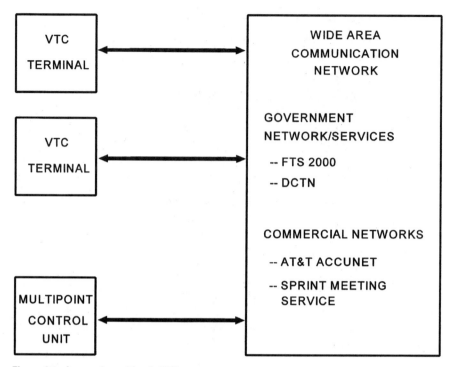

Figure 2.1 A generic multinode VTC system.

in complexity, ranging from a large multiroom configuration down to a desktop videophone device. In general, the transmission bit rate varies with the complexity. Large complex rooms used by many people and executive level personnel employ high-transmission bit rates (for example, 1.544 Mbps and 768 Kbps). Intermediate-sized rooms used typically for project work will employ intermediate bit rates (for example, 384 or 128 Kbps). Desktop videophones will employ the lowest rate (for example, 64 and 128 Kbps).

Although point-to-point VTC connections are very important, multipoint connections have been found to be even more important. Multipoint permits a large number of terminal nodes to simultaneously participate in a conference. Multipoint connections are accomplished by each terminal connecting to a multipoint control unit (MCU).

VTC systems are conveniently classified into four categories as defined by their physical configuration: customized conference room, rollabout module, TV set top, and desktop. Table 2.1 summarizes the key characteristics of these four different types of systems.

In general, if the VTC requirement calls for a number of people at a facility to routinely take part in remote conferences, there is a need for a customized

Table 2.1
Physical Configuration of VTC Systems

VTC Configuration	Physical Layout	Typical Display	TV Camera(s)	Lighting, Acoustics	Typical Scene; Number of People	Microphone/ Speaker
Customized room	Large conference room, large table, possible additional chairs. Electronic equipment is in a back room.	Rear screen projector(s) or multiple large TV monitors built into a wall.	Usually multiple cameras with pan, tilt, zoom.	Usually customized.	Large group of people; can be more than six.	Multiple microphones or one table top unit.
Rollabout	Self contained module(s), table within a conference room.	One or two large TV monitors built into rollabout module; small TV window for self view.	One or two cameras with pan, tilt, zoom.	Usually normal room lighting and acoustics.	Small group of people; up to six.	Typically one table top unit.
TV set top		One large TV monitor.	Integrated into the TV set top unit	Normal room lighting.	Small group.	Typically one table top unit.
Desktop	Camera/monitor on desktop; electronics in videophone or on floor. Typically a PC.	One small TV monitor. PC display.	One small camera. No pan, tilt, zoom.	Usually normal room lighting and acoustics.	Talking head; head and shoulders; usually one person.	Handset or integrated into videophone.

room or a room with a rollabout to accommodate people and to provide a convenient mechanism for integrating graphics and documents into the conference. If there is a requirement for a large number of people in the room (more than six) and if the personnel taking part in the conference are very senior, it may be necessary to implement a large customized conference room. If, however, the typical number of people using the room will be six or less, it is usually possible to place a rollabout modular unit in an existing room to minimize installation cost.

The key ITU standards and their function in an H.320 VTC terminal are as follows.

- H.320: Narrowband visual telephone system and terminal equipment;
- H.261: Video codec for audiovisual services at P x 64 Kbps;
- H.221: Frame structure for a 64 to 1,920 Kbps channel in audiovisual teleservices;
- H.242: System for establishing communication between audiovisual terminals using digital channels up to 2 Mbps;
- H.230: Frame synchronous control and indication signals for audiovisual systems;
- G.711: Pulse code modulation (PCM) of voice frequencies;
- G.722: 7 kHz audio coding within 64 Kbps;
- G.728: Coding of speech at 16 Kbps using low-delay code excited linear prediction (LD-CELP);
- H.231: Multipoint control unit for audiovisual systems using digital channels up to 2 Mbps;
- H.243: System for establishing communication between three or more audiovisual terminals using digital channels up to 2 Mbps.

2.4 Video Teleconferencing Room Considerations

2.4.1 Room Size, Furniture, and Location

The room designer must consider the types of meetings that will take place in the organization, such as small working group sessions, larger information-sharing meetings, or very large conferences, and select the most sensible video-conference room design for the organization. Before searching for a location for a videoconference room, it is necessary to know how much space is required, which depends on the number of people to be accommodated and the type of equipment to be used.

Let us begin with the number of on-camera participants. Each person should be allowed 3 ft of space along one side of the conference table. Thus, if the room is designed for six active conferees, an 18 ft-long conference table should be provided. In addition, there will need to be at least one aisle at one end of the conference table, which adds another 2.5 ft per aisle to the width of the room. We are now at 20.5 to 23 ft.

If a graphics camera is required, space must be allocated. If a separate cabinet is envisioned, another 2.5 × 2.5 ft of floor space will be required. Since many users prefer to have this capability near the conference table, the designer should consider allowing space for the unit in calculating the width of the room. Of course, ceiling-mounted graphics cameras do not require additional floor space.

There may be a requirement for additional off-camera seating. If the space under consideration is shaped to accommodate this area along the side, it will require another 2.5 ft per row of chairs and 2 ft per aisle between rows. Therefore, we have a room of at least 21.5 ft—ideally 25.5 ft or more—in width.

In determining the depth of the room, we must begin with the distance from the camera to the center of the conference table, which is typically 10 ft. The seats will take up another 3 ft when occupied, and an aisle of 3 ft should be provided. Therefore, the minimum distance from the camera to the rear wall is 16.5 ft. Some designs allow for extra seating behind the active participants rather than off to the side. A layout of a typical customized conference room is shown in Figure 2.2.

If a rollabout cabinet will house the room equipment, another 2 ft must be provided to account for the depth of the cabinet. There should be 6 in of clearance from the rear wall to facilitate air flow. If the installation sets the equipment into the front wall, there will need to be an equipment room to allow access to the equipment; allow at least 6 ft for this purpose. Therefore, a room with a cabinet system will need to be approximately 19 ft long, and a site-built facility that accommodates the same number of participants will need to be 22.5 ft. Of course, it is necessary to consider doors in the room layout, that is, doors for access to both the conference room and rear equipment room.

It is desirable to keep the ambient noise level of the room at a low level (45 dB or less). This is achieved by a combination of site selection and/or sound treatment of the walls. Room reverberation time should be between 0.3 and 0.5 sec. The room's absorption coefficient should lie between 0.25 and 0.45. If less than 0.25, the room will be hollow sounding; if more than 0.45, it will sound dead.

Most installations have an oval table with on-camera participants seated on one side of the table. Usually, the arc of the oval is designed so that every participant is equidistant from the monitor and camera. Everyone can see equally

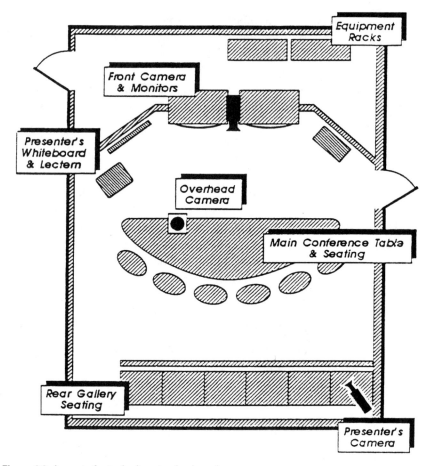

Figure 2.2 Layout of a typical customized conference room.

as well, and the only size differences apparent to the people viewing a distant monitor are those that actually exist. In this configuration, the people at another location are electronically on the other side of the table. This "we against them" seating arrangement may create an artificial barrier to effective group dynamics. If this is a major concern, another table shape should be considered.

A trapezoidal or blunted triangular table can be used to allow for two or three people on each of the two equal sides, creating the illusion of everyone seated at a square table, with four equal sides and no clear leadership position. A drawback is that those seated closest to the camera will appear larger than those farther away from the camera. The chairs selected should not have visually reflective parts that would cause the room lights to be reflected directly into the camera.

It is important to carefully locate the conference room within the facility. If the intended users are all top executives, it would be desirable to locate the room near the executive office area. If a mixture of senior level managers and other employees are expected to use the installation, it would be better to locate it in a neutral location. Try to avoid putting the videoconference room in an obscure part of the building. If possible, the location should not be adjacent to heavily traveled, noisy corridors. Outside walls and space adjacent to elevators shafts or equipment rooms should also be avoided.

2.4.2   Cameras

Generally, a camera can frame three seated participants and capture enough detail to communicate facial expressions or subtle body motions. For this reason, a typical room design with a six-person table requires two cameras, each viewing three people.

Since all VTC systems transmit only one video signal, means must be provided to integrate the two camera signals into the one video transmission channel. This is accomplished by either switching between the two cameras or using a split-screen technique. Camera switching can be initiated either manually or by voice activation. Voice activation is somewhat preferred because the picture of the speaker is automatically transmitted. Manual switching can be greatly simplified by using "presets" where a number of specific camera orientations (for example, pan, tilt, zoom, and focus) are stored in the system. A single operator command causes the camera to move quickly to a preset location.

Until now, we have only discussed cameras that view the participants, that is, "people cameras." In most VTC facilities, there is also a requirement to handle some visual aids such as flip charts, slides, transparencies, or solid objects. A separate camera is generally provided for this requirement. In some systems, this document camera is mounted in the ceiling or on a stand that can be placed on the conference table.

In rollabout systems, a separate graphics cabinet is often used. This cabinet may house a camera, a slide projector, a light table for transparencies, and a mounted lighting surface for capturing documents or small objects.

The ability to focus a camera on either a whiteboard or a podium is required in some installations. A single camera with a preset view or a separate camera is typically used for this requirement.

2.4.3   Display System

Early videoconference room designs used video projection units instead of monitors because large monitors were not available. There are several problems with projection systems, not the least of which is the high cost. The image is not sharp and clear due to the enlargement. The image is duller than monitor displays and must be viewed straight-on to avoid a further reduction in brightness. Finally, the camera is far from the center of the screen. In summary, video projectors are rarely used in VTC facilities; they are used only where there is a very large number of participants.

Today, monitors as large as 35 in (diagonal dimension) produce bright, clear images. The correct monitor size is determined by the distance between the monitor and the main participants. The general rule is that the height of the monitor should be one-eighth the distance between the screen and the viewer. In other words, if the participants are seated 10.5 ft from the monitor, the monitor height should be 15.75 in and the diagonal size of the monitor should be 26.25 in. As this is not a standard monitor size, monitors of either 25, 27, or 30 in may be used with confidence. A 35-in monitor would be the upper limit for a room this size. The center of the monitor screen should be approximately 3.5 ft above floor level.

Many VTC display systems being installed today use two TV monitors—one for "people pictures" and one for graphics. The scene viewed by the local camera is usually presented as a picture-in-picture (small window) on the people picture monitor. It is also common to implement a VTC system using a single monitor particularly with a fully packaged rollabout as shown in Figure 2.3. In such a system, it is obviously necessary to switch the function of the monitor between people picture and graphics or to use a window for the people pictures. Again, the local camera screen would be viewed using a small window.

2.4.4 Audio

Audio is the fundamental cornerstone of any teleconference system; graphics or video, although very important in most cases, must be considered to be supplementary. Unfortunately, there are many characteristics of an audio teleconference system that are very different from a conventional telephone system. Consequently, much of the telephony technology that has been developed over the decades is not applicable to the audio teleconferencing problem. Technical challenges that are unique to audio teleconferencing include room acoustics, microphone/speaker placement, echo cancellation, and bridging for multipoint operation.

The acoustic characteristic of the teleconference room can significantly affect the quality of the audio system. Ambient noise, acoustic reflectivity of the walls, and reverberation time are major factors contributing to audio quality. The technical characteristics and placement of the microphone(s) and speaker(s) in the room are also important factors in the performance of the

Figure 2.3 Front view of a rollabout unit with one monitor.

audio system. These factors will determine the tendency of a system to "howl" from feedback or generate an echo that must be canceled or suppressed.

Many rooms contain multiple microphones distributed around a conference table. The audio from these microphones can be combined in three different ways to form one signal for transmission. They can be mixed, manually switched, or automatically switched according to voice activity.

An objectionable echo can occur when voice signals are sent back to the originating end as a result of speaker-to-microphone coupling. The echo problem is solved either by suppression or cancellation. Echo suppressers automatically reduce the level of the return audio path when a local participant is talking, thereby reducing or eliminating the associated echo. Echo cancelers generate a pseudo-echo based upon the incoming audio signal and subtract this pseudo-echo from the transmitted signal containing the echo. In general, echo cancelers are more complex than echo suppressers and typically provide higher voice quality. Echo suppressers can exhibit a voice clipping characteristic that in some cases can be disturbing.

In the case of multipoint teleconferencing, an audio bridge is used to mix the voice signals originating from the several conference locations and distribute the output mixed audio to the conferees. The audio bridge must also control the magnitude of the output composite signal to avoid system overload.

In summary, the audio portion of a teleconference is absolutely crucial to the success of the conference. It is a technical challenge to design a system that provides high-quality audio under all conference circumstances.

Recently, an integrated audio product has appeared on the market from several vendors that shows promise of significantly easing the VTC audio problem. The device integrates microphone, speaker, and echo canceler into a single device that sits in the center of the table. This unit has promise for minimizing the need to acoustically treat the room and to be concerned with placement of microphone and speaker.

2.4.5   The VTC Controller

The remaining piece of equipment to discuss is the VTC controller. It is the central nervous system of the installation. The user controls and interacts with the equipment via the control unit. The control system must be user friendly. It should have the capacity to be expanded to meet the evolving needs of the organization, and of course, it must be reliable.

Control units can be divided into the two broad categories of wired or infrared.

2.4.5.1   Wired Control Units

Typically, wired units are a custom-designed array of mechanical controls such as buttons, switches, sliding potentiometers, and joy sticks. The front panel is custom inscribed to indicate control functions. Another popular design uses a touch-sensitive screen. Here again, the display can be customized to make the controller's operation easily understood by users.

All of these control units require a cable to connect them to the equipment room or rollabout cabinet. For this reason, wired control units are difficult to move around the conference table. The stiff cables and bulky units do not lend themselves to shared control of the meeting. In fact, some designs have the control unit built into the table, which is a limitation in an organization that encourages participation among meeting attendees. The person or people seated adjacent to the control panel become the meeting leaders.

If a podium or whiteboard is included in the installation, some of the controls will need to be duplicated and installed near the presenter. Also, the cable can be a physical hazard and must be installed in a duct. These wiring requirements add to the system's cost, and in the case of a rollabout, the transportability is reduced.

On the plus side, there is no space limitation on the face plate of a wired control panel. The designer may make the controls easy to use and very explicit. The type of control favored by the majority of users may be installed. For example, a slide control may be favored for the volume, while a joy stick may be easier to use when controlling the cameras. Other controls may be added to an existing wired controller as long as spare cable capacity was provided initially. Figure 2.4 illustrates the layout of a typical room controller.

2.4.5.2 Infrared Control Units

Infrared units may be either hand-held or table top devices. Therefore, the number of control functions is limited. System engineers often get around this constraint by giving some buttons multiple uses. The limited space on the face of the controller also restricts the amount of information that can be inscribed on the face plate. Users definitely require training to use these controllers. On a hand-held unit, only buttons are available. Other options such as slides or joy sticks cannot be easily implemented.

Table top infrared units can utilize more options. However, they are limited to a finite number of codes, with a separate code required for each function. Space is not a restriction. The face plate may be as explicit as the designer deems necessary.

There are also problems of lesser concern with this approach. All infrared units require batteries. These can begin to malfunction during a meeting and cause a great deal of user frustration.

The major strength of an infrared design is the ease with which the control of the meeting can be shifted. Not only can the unit be passed around, but

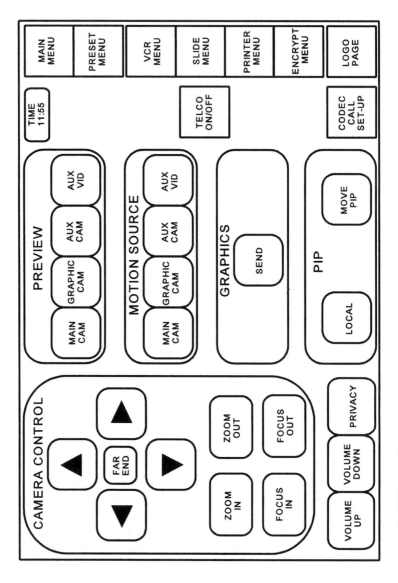

Figure 2.4 Layout of typical VTC controller.

there may be more than one unit in the room. For example, a separate controller can be placed at the location of the presenter and used when appropriate.

2.4.6 Administration

It is necessary to allocate significant personnel resources for the administration of a VTC room. The work activities of a VTC room administrator or coordinator are as follows:

- *Scheduling of conference calls:* This work can become complex, particularly to schedule multipoint conferences for a busy conference room. Computerized systems are available today to assist in the scheduling process.
- *Training:* There is a continuous need to train users, operators, and maintenance personnel.
- *Promotions:* It has been clearly established that a conference room is more productive if the use of the room is promoted within the user's community.
- *Operation:* Some complex VTC room systems require the use of a dedicated operator to support the users.
- *Maintenance:* The room coordinator must arrange for periodic and on-call maintenance of the room equipment.
- *Security:* Some conference rooms must handle encrypted calls. In these cases, considerable administrative effort is required to support the encryption equipment and related data.

2.5 Quality and Performance Considerations

2.5.1 General

A picture is worth a thousand words. This old saying not only gives the justification for teleconferencing but also states the fact that a picture contains much more information than the spoken word. Transmission of such information between distant locations requires an electronic highway that is wide enough to handle all required information. Greater width translates into higher cost. Therefore, in the past, pictorial transmission could be economically justified only for high-value applications such as TV broadcasting. Even though recent developments in transmission technology have considerably reduced this cost factor, it is likely to remain important for many years to come. Video teleconferencing has become economically viable only by using special techniques that drastically reduce the amount of required transmitted information.

Fortunately, every picture contains a large amount of redundancy. There are always areas that do not contain details and, thus, do not call for the transmission of a large amount of information. Even with a large amount of detail, if the detail does not change rapidly, very little information must be transmitted. Furthermore, even in the case of a fast moving scene, the changes at any one point are limited and often predictable. Thus, it is not always necessary to transmit a complete picture. Limiting the transmitted information to changes is generally sufficient to reconstruct the picture at the receiving location. Applying such techniques can result in a large amount of bandwidth compression or data rate reduction.

Unfortunately, everything has its price. Most compression techniques result in at least some loss of picture quality. The development of more advanced techniques makes this loss less and less noticeable. Furthermore, a teleconference normally does not call for the high picture quality that we are accustomed to in entertainment TV broadcasting. This makes the application of compression techniques both technically and economically viable.

2.5.2 System Considerations

Video teleconferencing is still a relatively new technology. Many manufacturers in the United States, Japan, and Europe have developed techniques for video compression. In the past, most of these techniques were proprietary, meaning that equipment of different manufacturers could not operate with each other. In the beginning, this was no problem since teleconferencing systems were small and independent of each other. However, after a few years of development, it became obvious that this lack of interoperability presented an intolerable constraint, not only within the United States but even more so in global applications where different TV standards (such as NTSC, PAL, and SECAM) have to be accommodated. Therefore, the ITU developed a standardized technique (algorithm) for digital video encoding, Recommendation H.261, which can be made compatible with all national TV broadcast standards. Of course, equipment with proprietary algorithms may remain in use for some time to come, but for future systems only equipment conforming with H.26X and its associated standards need and should be considered.

2.5.3 Picture Format

None of the existing analog TV broadcast standards is suitable for compressed video transmission. Compression techniques call for a digital signal, which requires breaking down the picture into individual samples or elements (pixels). Most important information is contained in the monochrome or luminance portion of the picture; therefore, transmission is performed in separate luminance and chrominance (or color difference) components at different levels of quality. Standard sync and blanking signals can be regenerated in the receiver and need not be transmitted, thus using all the available transmission channel for picture information.

2.5.4 Common Intermediate Format

The H.261 standard defines a common intermediate format (CIF) as seen in Table 2.2. The best obtainable resolution is roughly half that of the NTSC broadcast standard, which is roughly comparable to the resolution provided by a consumer-quality VCR. This resolution is suitable for group conferences, where each participant takes up only a small fraction of the viewing area. However, if the available bit rate is low, the result may be jerky motion. For graphics, CIF can be used to transmit simple viewgraph text images that have about 15 or fewer lines of text. Note also that the CIF is not required of all codecs, so conferences with other codec models may not be able to operate at CIF resolution. CIF is recommended for all codecs that will be operated at bit rates of 384 Kbps and above.

2.5.5 Quarter Common Intermediate Format

The lower quality quarter common intermediate format (QCIF) is limited to only half the resolution of CIF in each dimension and, therefore, has one quarter the number of pels as CIF. Clearly this resolution is noticeably poorer than

Table 2.2
H.261 Picture Format

Parameter	Luminance Y	Chrominance R-Y	B-Y
Full CIF			
Pels/line	352	176	176
Lines/picture	288	144	144
Quarter CIF			
Pels/line	176	88	88
Lines/picture	144	72	72
Picture Rate: Submultiple of 29.97 Hz			

entertainment quality TV. This resolution is suitable for single-person conferences, where only the head and shoulders of one person are shown. For graphics, QCIF can be used only for very simple viewgraphs that have at most seven
lines of text. All H.261 codecs are required to have QCIF, even if they have CIF
capability, so it provides a fall-back mode that will work between all types of
codecs. Even if a codec has CIF capability, most codecs provide a feature that
forces operation at QCIF if it is desired to improve frame rate or to economize
on bit rate.

2.5.6 Frame Rate

The TV broadcast standard of 30 frames per second produces a perfectly
smooth picture. This is important for entertainment but not for teleconferencing. Therefore, a very effective way to reduce the transmitted bit rate is not
to transmit all frames and "fill in" at the receiver by simply displaying the last
transmitted frame several times. The resulting received picture may not show
smooth motion. Indeed, it may appear as a series of snapshots. However, this
is adequate for many applications. Indeed, many observers find it preferable to
a blurry picture with smoother motion. The picture rate, though theoretically
allowed to be at the NTSC standard, 30 frames per second, will normally be
lower. This feature provides a wide range of freedom in reducing the transmission requirements. A value of 15 frames per second is still almost perfect; many
observers consider 10 frames per second acceptable for teleconferencing, but
lower values (which often occur at low bit rates) may look rather like a series
of still pictures than continuous motion. In most recent codecs, the transmitted frame rate is adaptive, meaning that the algorithm automatically tries to
optimize the combination of encoding accuracy and motion smoothness.
However, it is possible to manually select a minimum permissible transmitted
frame rate.

2.5.7 Motion Compensation

The above features produce a considerable but often not sufficient amount of
data compression. An added feature is motion compensation, which uses
frame-to-frame comparison of blocks that are not only in the same but also in
adjacent locations. This way it can be determined in which direction a part of
the picture has moved between frames. Motion compensation can significantly
improve the quality of the picture. H.261 requires implementation of motion
compensation in the decoder section of the codec, but it is optional in the encoder.

2.5.8 Bit Rates

The available (or affordable) transmission facility for a teleconference system generally determines the bit rate at which the codec can operate. This rate is manually selected. The higher the bit rate, the less data compression is needed and the higher the quality of the viewed picture.

The transmission bit rates are specified in terms of P × 64 Kbps. Based on commonly available transmission facilities, the highest value of P is usually 24 in the United States and 30 in Europe. P = 1 corresponds to a single B channel in ISDN or a switched 56 channel. P = 2 corresponds to an ISDN basic rate interface (BRI) or a pair of switched 56 channels. P = 6 through 12 correspond to fractional T1 circuits, while P = 24 is a full T1.

The resolution obtainable with QCIF is so limited that using a higher rate would be a waste of resources. A value of P = 1 is possible but can be used in practice only for a very low-quality picture and with G.728 audio. In practice, some transmission capacity must be reserved for audio and other ancillary functions. P = 1 is marginally usable with CIF if there is little motion in the picture. At P = 6, CIF provides acceptable picture quality for most video teleconferencing applications. Higher rates (P = 12, 24) produce pictures that, though still below broadcast quality, are judged only slightly degraded by most observers.

The state-of-the-art in video compression technology is in a steady flux, and new developments result in ever-improving performance. This allows the use of lower and lower bit rates not only for videophone but for many other video teleconferencing applications.

2.5.9 Audio

Audio is a vital part of any teleconference. Indeed, it is generally considered more important than video since its lack would destroy the usefulness of the conference.

Furthermore, audio teleconferences have been around for many years and participants have become accustomed to excellent quality. Therefore, an adequate portion of the available transmission bit rate must be reserved for audio.

There is no single audio transmission standard assigned for use with the H.261 video codec. Table 2.3 lists three available choices. The G standards are officially established by the International Telephone and Telegraph Consultative Committee (CCITT). With a low available total bit rate it is obviously important to keep the requirement for audio as low as possible. At 384 Kbps or above, it is probably feasible to assign 48 or 64 Kbps to audio, resulting in high quality without imposing an undue constraint on video.

Table 2.3
Audio Coding Standards

ITU Recommendation	BW	Bit Rate	Coding Algorithm
G.711	3 kHz	64 Kbps	PCM
G.722	7 kHz	48, 56, 64 Kbps	Dual band, DPCM
G.728	3 kHz	16 Kbps	LD-CELP

2.5.10   Bandwidth

The wider the audio bandwidth, the more intelligible and natural speech becomes. The 3-kHz bandwidth shown in Table 2.3 for G.711 and G.728 corresponds to toll quality, similar to that experienced in normal analog telephone service, and provides good intelligibility. The 7-kHz bandwidth for G.722 is higher quality and will provide more natural sounding speech but is not as good as FM radio or CD quality. Most users of VTC will definitely prefer the higher quality audio.

2.5.11   Lip Sync

An important characteristic of the digital audio system is its inherent transmission delay. The different processing to which video and audio signals are subjected basically produces different delays. For a video teleconference close lip sync is desirable. Any noticeable deviation can be extremely annoying to all participants. Therefore, the processing delays of the video and audio channels should be accurately equalized.

2.5.12   Echo Cancellation

Many rooms contain multiple microphones distributed around a conference table. The audio from these microphones can be combined in three different ways to form one signal for transmission. It can be mixed, manually switched, or automatically switched according to voice activity.

An objectionable echo can occur when voice signals are sent back to the originating end as a result of speaker-to-microphone coupling. The echo problem is solved by two different approaches: suppression or cancellation. Echo suppressers automatically reduce the level of the return audio path when a local participant is talking, thereby reducing or eliminating the associated echo. Echo cancelers, illustrated in Figure 2.5, generate a pseudo-echo based upon the incoming audio signal and subtract this pseudo-echo from the transmitted signal containing the echo. In general, echo cancelers are more complex than echo suppressers, and they typically provide higher voice quality. Echo suppressers can exhibit a voice clipping characteristic that in some cases can be disturbing.

The codec used for teleconferencing provides for continuous audio transmission in both directions. This is called full duplex. However, with a typical room configuration the sound from the far end of the room is radiated from the room loudspeakers and finds its way, either directly or by means of reflections, to the microphones. This causes an echo to be heard in the far end of the room of the sounds originating there. This is a very annoying effect and should not be tolerated. There are three basic solutions to this problem.

1. The simplest is to use a conventional telephone handset. The configuration of the handset effectively decouples the loudspeaker and microphone. However, this approach is feasible only for single-user videophones and even then is not desirable.

2. Another approach is to sense in which room speaking is actually taking place at any instant of time and only allow sounds to be transmitted from that room. Since this approach allows transmission in only one direction at a time, it is known as half duplex, or echo gating. This technique is widely used in speakerphones. However, it is not completely satisfactory for VTC because of the increased delay

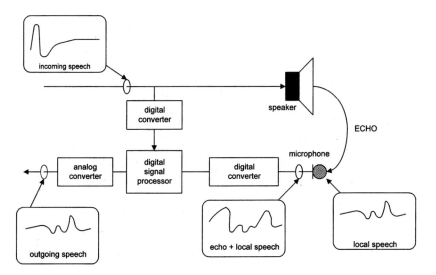

Figure 2.5 Acoustic echo cancellation.

involved. Conversations are awkward and whole words or phrases may be completely lost due to switching.

3. By far the best approach is to use echo cancellation. In this technique the incoming and outgoing audio signals are compared electronically, and the portion of the outgoing signal that is due to the incoming signal is subtracted from it, ideally leaving only that portion of the signal that actually originated in the near end of the room. While cancellation is not perfect, most echo cancelers do a very good job.

2.5.13   Error Control

Digital transmission of audio and video signals is relatively tolerant to transmission errors, but transmission errors still are an important cause of audible and visual defects. There is usually a rather sharp error threshold between good performance and failure. With interframe coding, the effect of uncorrected errors can persist for quite some time.

H.261 requires the addition of the FEC bits in the encoder but does not mandate their use in the decoder. Since the bits have already been used in the transmission, it makes sense to use them in the decoder. How much FEC will improve performance depends on the bit rate, the error rate, and the type of errors produced by the network. For random bit error rates of about 10-8 or better, the picture will appear virtually error free even without FEC. Between 10-8 BER and 10-3 BER the FEC will improve performance but will be quite poor at 10-4. At error rates poorer than 10-3 the system will probably be unusable.

2.5.14   Performance Impairments

The various techniques available to achieve a large amount of data compression inevitably result in an impaired picture. The user has to determine what amount and type of impairment is tolerable for a particular application of the teleconferencing system. This will provide guidance for the selection of terminal equipment and transmission facilities. The following is a listing and brief description of the major impairments.

- *Blurriness:* There are two types of blurriness. One is static and is mainly caused by limited resolution due to the picture sampling process, particularly with QCIF. Limitations imposed by the intraframe encoding procedure also add to this impairment. The other type is best described as smearing of moving objects. It is caused by limitations in the interframe coding, which makes it impossible for the encoded picture to follow changes with sufficient speed.

- *Blocking:* This impairment manifests itself as a spurious checkerboard pattern superimposed on the picture. It is caused by imperfections in the encoding of the 8 by 8 pixel block structure specified by the H.261 algorithm.

- *Image retention:* Similar to the smearing of moving objects, interframe coding often does not allow fast enough updating of a sudden picture change. This shows up, for instance, as retention of a pattern after it has been erased or a noticeable delay in updating the image after a switch between radically different pictures.

- *Jerkiness:* Any motion in the picture lacks smoothness but appears as a sequence of jumps. This is caused by frame repetition, which results in a lowered average transmitted frame rate.

- *Quantizing noise:* The quantizing process results in discrete steps that do not accurately represent the input. The difference produces a spurious signal component that appears similar to conventional noise. Generally, this impairment is only slightly noticeable.

Other types of picture impairments are often mentioned, but they are generally used in an analysis by experts. In both appearance and cause they are fairly similar to one or more of the aforementioned impairments.

Picture Quality Is Not Standardized

The H.26X video coding standards are absolutely necessary to achieve interoperability between multimedia terminals from different manufacturers. However, adherence to video coding standards does not guarantee the quality of the output TV signal. Figure 2.6 illustrates what *is,* and is not, standardized by a video coding standard. It shows that the transmitted bitstream and the decoder *are* standardized while the encoder, pre-processor, and post-processor are not. Unfortunately, the decoder, which *is* standardized, does not determine the quality of the output video signal. The decoder, which must rigorously follow the instructions of the encoded bitstream, has little intelligence. Quality is completely determined by those parts of the digital TV system that are *not* standardized. Picture quality is primarily determined by the encoder, but manufacturers are continually introducing new pre- and post-processing technology to significantly improve the output picture. Error concealment is one important example of a post-processing technique that is now being widely deployed. As the encoder codes the input picture, it continually makes very complex discretionary decisions that fundamentally determine picture quality. Examples of these proprietary encoder actions are (1) the selection of picture blocks to be transmitted, (2) transmission mode, (3) quantization level, (4) intra/inter decision,

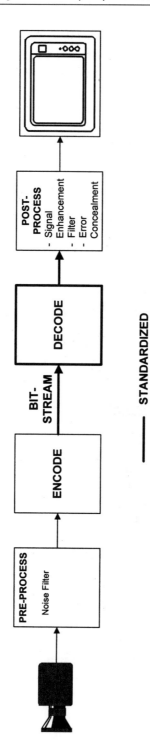

Figure 2.6 Digital TV standardization.

(5) buffer control, (6) frame rate, (7) motion tracking algorithm. In summary, it is the encoder, not the decoder, that determines picture quality. Because the fundamental encoding decisions are not standardized, video encoders from different manufacturers inevitably deliver different video quality.

2.5.15 Standards To Measure the Quality of Digital TV Systems

The ANSI and ITU standards organizations are beginning to address the very difficult problem of objectively measuring the performance of digital television systems. The problem is by no means solved at this point in time, but a few of the standards that have been adopted to address the issue are described below.

2.5.15.1 ANSI T1 Committee Standards

ANSI T1.801.01—Test Scenes

This standard defines a set of 25 test scenes that were assembled by the ANSI/T1A1.5 Committee for use in the development of techniques to assess the performance of digital video teleconferencing and videophone systems. These scenes were used in subjective tests and will be used in future objective tests. The purpose of the standard is to make available to the industry a collection of video test scenes to aid in the development of objective test methodologies that statistically correlate to the subjective performance.

The video scenes have been grouped according to five scene-content categories: (1) one person, mainly head and shoulders—four scenes; (2) one person with graphics and/or more detail—seven scenes; (3) more than one person—six scenes; (4) graphics with pointing—five scenes; (5) high object and/or camera motion—three scenes. A single image from each of the test scenes is included in the standard.

All of the video test scenes described in the standard are in the public domain. Copies of a videotape containing the test scenes are available from ANSI. An annex describes the timing of the test tape in detail. The typical duration of one test sequence is 13 seconds.

ANSI T1.801.02—Performance Terms, Definitions, and Examples

This standard specifies a set of terms and definitions that are applicable to the digital transport of video teleconferencing and videotelephony signals. The purpose of the standard is to define a common terminology for use in the VTC/VT community, thereby improving communication among current and future members. A total of 15 general terms (e.g., lip sync, scene cut, spatial performance) and 16 impairment terms (e.g., blurring, edge busyness, jerkiness) are defined.

A set of illustrative video clips is included with this standard, primarily focusing on the impairment terms but illustrating some of the general terms as well. The tape, which is available from ANSI, is over 10 minutes in length, and contains 26 examples of terms. When illustrating an impairment term, usually both the unimpaired and impaired versions of the video clip are present on the tape for comparison.

ANSI T1.801.03—Digital Transport of One-Way Video Signals—Parameters for Objective Performance Assessment

This standard covers the operational assessment of one-way 525-line video systems utilizing digital transport facilities. It gives the measurement parameters that may be used to detect changes in the current status of a system when used in comparison with a set of reference measurements on the same system made under initial provisioning circumstances. Additionally, there are diagnostic parameters identified in this standard that may be utilized to characterize aspects of one-way video signals. Four objective parameters are defined using artificial test scenes, and 20 objective parameters are defined using natural test scenes.

2.5.15.2   ITU Standards

Internationally, the standardization work is performed by Study Groups 9 and 12 of the ITU-T. A list of the standards developed by these organizations is listed below and described in the following sections:

- P.910—Subjective Video Quality Assessment Methods for Multimedia Applications;
- P.920—Interactive Test Methods for Audiovisual Communications;
- P.930—Principles of Reference Impairment System for Video;
- P.AVQ—Subjective Assessment Methods for Global Audiovisual Quality Evaluation in Multimedia Application.

ITU P.910

This standard defines noninteractive subjective assessment methods for evaluating the quality of digital video images coded at low and medium bit rates (up to 2 Mbps) for applications such as videotelephony, videoconferencing, and storage and retrieval. The recommendation covers the following topics:

- Laboratory set up to produce test sequences;
- Laboratory set up to carry out subjective assessment;
- Characteristics of the test sequences;

- Test methods and experimental designs;
- Analysis of data.

ITU P.920

Recommendation P.920 defines interactive evaluation methods for quantifying the impact of coding artifacts and transmission delay on point-to-point or multipoint audiovisual communications. This methodology is based upon conversation opinion tests, and can be considered to be an extension of methods defined in ITU-T Recommendation P.80, Annex A. Three major topics are covered by the recommendation: (1) conversational tasks to be performed by test subjects, (2) methods and experimental design, (3) questionnaires.

Interactive audiovisual tests require the definition of a task to be performed by the test subject. The task must be as natural as possible. However, at the same time, it must stimulate the interactive communication, and the outcome must be somewhat quantifiable. Examples of tasks are name-guessing, story comparison, and picture-comparison. Several evaluation scales are offered to evaluate the performance of the task.

ITU P.930

Recommendation P.930 (entitled Principles of a References Impairment System for Video) describes the principles of an adjustable video reference system that can be used to generate the reference conditions necessary to characterize the subjective picture quality of video produced by compressed digital video systems. A Reference Impairment System for Video (RISV) can be utilized to stimulate the impairment resulting from the compression of video sequences, independent of compression scheme.

An RISV is capable of producing the following categories of distortion, either singly or in combinations, with independent adjustment of each impairment level:

1. Artifacts due to conversions between analog and digital formats (e.g., noise and blurring);
2. Artifacts due to coding and compression (e.g., jerkiness, edge busyness, and block distortion);
3. Artifacts due to transmission channel errors (e.g., errored blocks).

In this recommendation, five types of impairments (block distortion, blurring, edge busyness, noise, and jerkiness) are defined and general methods for implementing these impairments are provided.

From the viewer's point of view, the impairments produced by the RISV should be a good approximation of impairments generated by digital video coding and transmission systems.

Three possible applications for the RISV are: (1) creating reference conditions in subjective tests of digital video systems to ensure that the quality of the scenes presented to viewers covers the entire range of picture quality, (2) defining standard video impairment levels that can be used to compare subjective test results, and (3) quantifying the user-perceived quality of a video system with respect to a known reference.

Although this recommendation describes the principles of an RISV, before an implementation can be recommended, validation tests are required.

ITU P.AVQ

Draft Recommendation P.AVQ (entitled Subjective Assessment Methods for Global Audiovisual Quality Evaluation in Multimedia Applications) defines noninteractive subjective assessment methods for evaluating the global quality of digital video with accompanying audio for audiovisual applications such as videotelephony, videoconferencing, and storage and retrieval.

The recommendation describes the methods which are suitable for evaluating the following quality aspects:

1. The global effect of coding impairments, taking into account the interactions between audio and video degradation;
2. Problems related to lack of synchronization between the two signals;
3. The impact of transmission errors on such video systems.

These methods can therefore be used for several different purposes including, but not limited to, selection of algorithms, ranking of audiovisual system performance, and evaluation of the quality level during an audiovisual connection.

The recommendation describes test methods and experiment design, evaluation procedures, and statistical procedure for reporting results.

2.6 Planning a VTC System

This section provides the reader with some guidance on the techniques and procedures to plan for the introduction of a VTC system. We discuss steps in VTC analysis such as user survey, scheduling, and communication.

2.6.1 User Survey

The first step in planning a VTC room or multisite system is to survey potential users of the VTC system. The primary purpose of the survey is to determine the amount of travel used by personnel in the organization. In general, organizations whose personnel travel extensively will benefit most from VTC. The survey would be implemented using a questionnaire that would obtain:

- Number of trips per year;
- Number of persons per trip;
- Trip duration;
- Level of participants;
- Destinations.

2.6.2 Develop a Technical Approach

The second step in the planning process is the development of the technical approach for the implementation of the VTC system. Decisions that must be made are:

- Room size, which depends largely on the typical number of participants per conference;
- Electronic equipment, such as the number and size of TV monitors, cameras, graphics scanner, audio equipment, control system, and possible encryption;
- Transmission bit rate, which determines picture and audio quality;
- Custom room, rollabout, or desktop configuration;
- Multipoint requirements—if a multisite system is being developed.

2.6.3 Cost Estimate

The next step is to estimate the cost of the VTC system. The elements of the system for which cost data must be developed are nonrecurring, such as:

- Electronic equipment;
- One time communication access;
- Furniture;
- Room modification, if any;
- Installation.

Also consider the recurring costs, such as:

- Communication;
- Maintenance;
- Administration, training, and operation.

2.6.4 Schedule

An important part of the planning process is the establishment of a schedule for the VTC project. Important elements of the schedule are:

- Survey;
- Development of a technical approach;
- Cost estimate;
- Cost/benefit analysis;
- Finalization of the plan and proposal;
- Decision to (or not to) implement;
- Ordering materials;
- Ordering communications;
- Employment and training of staff;
- Installation;
- System test;
- User indoctrination;
- Project completed.

2.6.5 Project Proposal

Assuming the results of the cost/benefit analysis are positive, the final step in the planning process is to organize all the planning data into a document that describes the proposed project. An example of an outline of a proposal is as follows:

1. Executive summary;
2. Requirement analysis and survey;
3. Implementation plan;
4. Cost estimate;
5. Cost/benefit analysis;
6. Schedule;
7. Conclusions.

3

Video Compression Techniques

3.1 Introduction

The bandwidth of the basic analog TV signal is 4 MHz, which although well suited for broadcast TV, video recording, and closed-circuit TV applications, is very incompatible with the world of point-to-point interactive telecommunications such as videoconferencing and videophone. To enable videoconferencing and videophone to become economically practical, the video signal had to be digitized and compressed down to bit rates that are available in standard circuit-switched communication networks such as ISDN. And, since the cost and ubiquity of communication resources is much more favorable at low ISDN bit rates, there has been, and always will be, a strong motivation to reduce the video bit rate to the absolute minimum. A great deal of research and development has been devoted to the topic of video compression, and much success has been achieved. An uncompressed National Television Standards Committee (NTSC) video signal requires a bit rate of at least 90 Mbps. The most common video bit rate employed in videoconferencing systems today is approximately 100 Kbps, yielding a compression ratio of roughly 1,000 to 1. This incredible achievement in video compression has enabled the unfolding revolution of digital television as now implemented in systems such as videoconferencing, satellite TV, HDTV, DVD, and videophone.

This chapter provides an overview of video compression *technology.* It does *not* describe particular video coding standards which have been so important to the introduction and explosion of videoconferencing and videophone systems. These video coding standards are presented elsewhere in this book.

The first videoconferencing system standard, established by the ITU in 1990, is designated as H.320 and defines the means for communicating multimedia signals over the N-ISDN. The video coding standard for this H.320 system is H.261 and is described in detail in Section 6.3. The ITU has recently completed an improved video coding standard know as H.263. Although this new standard was developed primarily for transmitting video over very low bit rate channels such as POTS and the Internet, it is also applicable for higher bit rates such as N-ISDN. The H.263 video coding standard is described in detail in Section 8.4. The International Standards Organization (ISO) has developed the MPEG series of video coding standards which are primarily employed in the broadcast TV industry. The MPEG standards are presented in Chapter 11.

3.2 Overview

3.2.1 Basic NTSC Scanning Format

A single frame of a television image is comprised of 525 horizontal scan lines. If these frames were displayed at a rate of 30 frames/s, they would exhibit visual flicker. The frames could be displayed at twice that rate, 60 frames/s, but then the rate at which the information was being transmitted would double, thereby requiring twice as much bandwidth. Images fuse and appear flicker-free for most humans at rates faster than about 40 Hz. Cinema movies are shot at 24 frames/s. If they were displayed at a rate of 24 images per second, the flicker would be objectionable. The trick is to display each frame twice using a shutter rate of 48 images per second. A similar trick is used in television.

Each frame is divided into two fields with one field composed of the odd-numbered horizontal scan lines and the other field composed of the even-numbered horizontal scan lines. The field rate is twice the frame rate, or 60 fields per second, thereby avoiding visual flicker. If each field were simply displayed on top of the other field, however, resolution would be halved. The solution is to interlace the two fields so that the scan lines for the odd field fall exactly in between the scan lines for the even field. This technique is called interlacing and is depicted in Figure 3.1.

Interlacing preserves the vertical resolution of 525 scan lines and avoids visual flicker of large areas of the reproduced image. Flicker, however, *can* occur in a small area of the image. For example, if a scene with a horizontal line were scanned, then this line would appear only in alternate fields. It hence would be displayed at a rate of 30 images per second and would flicker. Luckily, most scenes are complex enough that small area flicker is not a problem in commercial broadcast television.

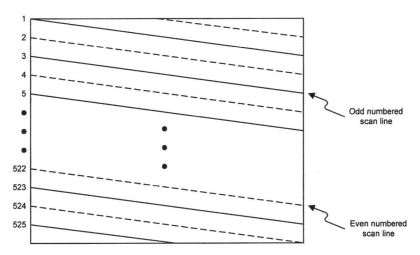

Figure 3.1 An NTSC video frame.

The total number of horizontal scan lines in a frame is 525 lines. Each frame is composed of two fields, and each field consists of 262.5 lines. Field 1 consists of all the odd-numbered lines in a frame, and field 2 consists of all the even-numbered lines. The first line of the odd field starts at the upper-left corner of the picture, and the last (263rd) line of the odd field is a half line that ends at the bottom half of the picture. The first line of the even field begins at the top half of the picture, and the last line of the even field ends at the bottom-right corner of the picture.

Time is needed for the scanning spot to travel from the bottom of the screen back to the top to begin the next field. This time is called the vertical retrace or blanking interval. The amount of time required for 21 scan lines is allowed in each field for the vertical retrace interval.

3.2.2 Compression Overview

Figure 3.2 is a functional block diagram of a generic system that digitally transmits video over a communication channel. At the transmitter, the input analog signal is first filtered such that the upper cut off frequency of the signal is N cycles/s. The filtered signal is next sampled at a rate of at least 2 N samples per second (the Nyquist rate) to avoid aliasing distortion. Each sample is defined as a pixel (picture element) that is commonly encoded with 8-bit accuracy because this precision is required to avoid any visible distortion in the output image. At this point the bit rate is typically 16N bps, which may exceed the bit rate of the transmission channel (C bps). The purpose of the compressor is to reduce the

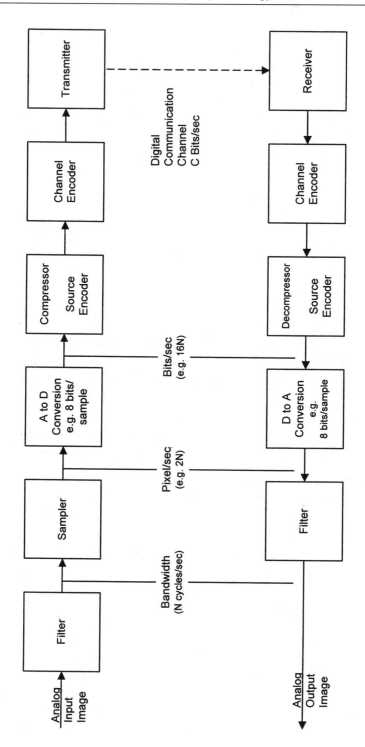

Figure 3.2 A generic system for the digital transmission of images.

16N-bit rate by reducing the pixel-to-pixel redundancy inherent in the image. The channel coder (for example, modem) processes the binary compressed signal for efficient transmission over the communication channel. The compressor is commonly referred to as a source coder (signal source) as contrasted with the channel coding process. As shown in Figure 3.2, the functions at the receiver are inverted relative to those at the transmitter.

The ITU H.261 video coding standard is universally used for virtually all video teleconferencing systems. The picture format most commonly used for videoconferencing is the full CIF (FCIF), which is defined in Table 3.1.

If the picture rate is 30 frames/s and each input pixel is encoded to 8-bit precision, the resultant bit rate of the uncompressed signal to be coded for transmission is 36.5 Mbps. Transmission bit rates for typical communication channels are 1.544 Mbps (a T1 channel) and 384 and 64 Kbps. The purpose of the video coder is to compress the input signal so that it can be transmitted through channels of this type. The required compression ratios for these transmission data rates are 24.3, 95, and 570 to 1, respectively. The compression problem is clearly a formidable one.

The required compression defined above uses the FCIF as a reference for spatial resolution. FCIF resolution is ideal for the videoconferencing application, but is inadequate for other applications requiring higher quality such as broadcast TV. The ITU has developed standard CCIR-601 to specify the resolution necessary to digitally encode a full-frame video signal with sufficient precision to provide the quality required for the broadcast TV application. There are two versions of CCIR 601: one for the NTSC video system and one for PAL. The NTSC version defines a spatial resolution of 720×480 pixels for the luminance signal while the PAL resolution is 720×576 pixels. CCIR-601 specifies a chroma resolution, known as 4:2:2, which is half the luminance resolution in the horizontal direction—360×480 and 360×576 pixels, respectively. If each of the pixels defined by the CCIR-601 signal is encoded with 8-bit precision, the resultant bit rate is 165.89 Mbps. Therefore, if the CCIR-601 signal is used as

Table 3.1
Pixels/Frame for CIF Picture

	Pixels/Line	Lines/Picture	Total Pixels
Luminance	352	288	101,376
Chroma R-Y	176	144	25,344
Chroma B-Y	176	144	25,344
			152,064

a reference to define the requirement for video compression, the required compression ratios for transmission bit rates of 1.544 Mbps and 384 and 64 Kbps are 107, 432, and 2,600 to 1, respectively.

3.2.3 Analog-to-Digital Conversion

Analog signals, originating from sources such as audio, video, and data sensors, are converted to a digital format by means of a quantization process. Figure 3.3 illustrates the operation of a basic scalar quantizer which linearly converts an input analog signal to an output digital form. The quantizer, having discrete decision points (e.g., 1, 2, 3), generates an output signal having restricted quantized output levels (1.5, 2.5, 3.5). Figure 3.3 illustrates a 2-bit quantizer dividing the signal dynamic range into four output levels. The typical quantization precision required for the coding of video and audio signals to avoid visible and audible distortion is 8 bits having a total of 256 levels of quantization. Signals are typically quantized to high precision prior to their being fed to a computer or codec for processing. Quantization is also a key element in the video compression

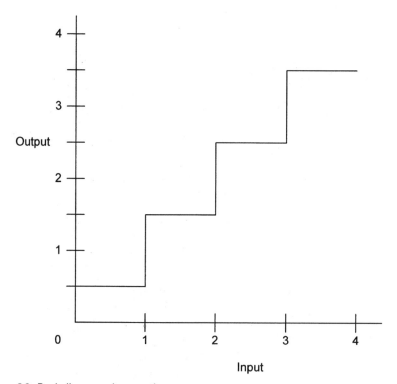

Figure 3.3 Basic linear scalar quantizer.

process itself. In this case, the quantization of the prediction error is dynamically varied to adapt to fluctuating picture content. Quantization is coarse in those regions of the pictures incurring rapid motion while quantization is fine when the picture is stationary.

3.2.4 Pulse Code Modulation

In PCM coding, the signal is sampled at the Nyquist rate, each sample is quantized to 2^m levels, and each level is represented by a binary word m bits in length. At the decoder, these binary words are converted to a series of amplitude levels that is low-pass filtered. Since each sample is encoded independently of its neighbor, no redundancy is reduced and PCM is used as reference to measure the performance of bit rate reduction techniques. The PCM coding precision that is typically used for this reference is 8 bits per pixel because this precision assures that there is no visible degradation in the picture. Figures 3.4, 3.5, and 3.6 illustrate the types of contouring distortion that occur when the coding precision is limited to 2-, 3-, and 4-bit precision, respectively.

The eye is logarithmic in its sensitivity to brightness distortion; that is, it is equally sensitive to equal percentage changes in brightness. For this reason the PCM quantization is typically nonlinear with the size of the quantizer steps

Figure 3.4 Two-bit PCM encoding.

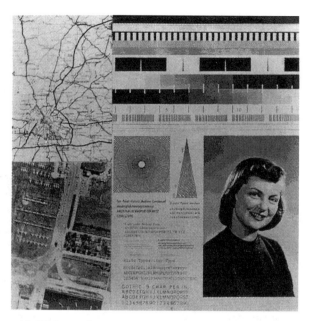

Figure 3.5 Three-bit PCM encoding.

Figure 3.6 Four-bit PCM encoding.

increasing with brightness. This nonlinear companding improves the picture quality for a given number of bits per pixel.

3.3 Video Compression Overview

The efficient coding of video signals has been the subject of research and development for many years, and a large number of different compression algorithms have been studied. The most well-known coding techniques of the past, present, and future are listed in Table 3.2. The first two—PCM and predictive (differential PCM (DPCM))—are very basic coding techniques. PCM provides no compression and is used as the reference to measure the performance of coding techniques. DPCM is still used today for the lossless coding of still pictures (for example, JPEG). Transform coding is the most dominant technique used to transmit video today. The discrete cosine transform (DCT) is used in the H.261, MPEG, HDTV, and H.263 standards. The next three techniques—vector quantization, wavelets, and fractals—can be considered waveform coding techniques that are competitive with the DCT. Finally, object-based coding is more advanced than the others since it isolates and encodes objects in the scene. Table 3.2 summarizes three key characteristics of these techniques, and each of these coding techniques is discussed in turn.

Figure 3.7 is a functional block diagram of a generic video compression system. It shows that the compressor is typically implemented in three sequential operations—signal analysis, quantization, and variable length coding. Basically,

Table 3.2
Coding Technique Overview

Coding Techniques	Geometric Element of the Picture Being Encoded	Content Based	Symmetry*
PCM	Pixel	No	S
Predictive (differential PCM)	Pixel	No	S
Transform	Square block of pixels	No	M
Vector quantization	Square block of pixels	No	MM
Wavelet	Multiresolution filtered elements	No	M
Fractal	Block having any size or shape	Yes	MM
Object-based coding	Moving objects	Yes	MM

*Complexity of the encoder relative to the decoder: S, essentially the same; M, more complex; MM, much more complex.

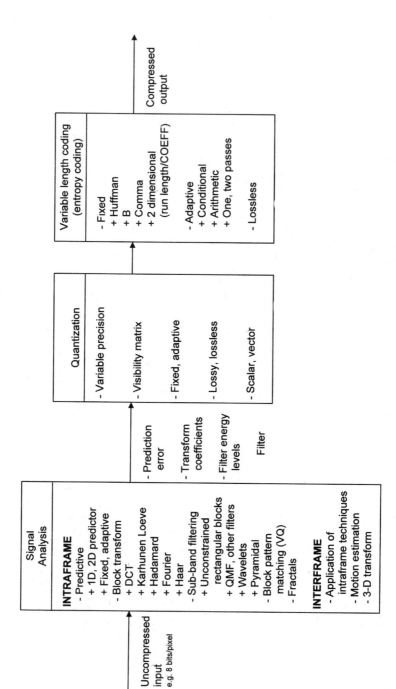

Figure 3.7 Functional block diagram of a video compression system.

the signal analyzer performs measurements on the input uncompressed pixels. For example, the analyzer may compute prediction errors, compute transform coefficients, filter the signal into subbands, correlate the pixels with prestored VQ patterns, or correlate the pixels with the image itself (fractals). The analyzer performs measurements within a single frame (intraframe) and/or from frame to frame (interframe). Typically, no compression is achieved by the signal analysis function. The input pixel data is merely transformed into another format that is more compressible than the original signal format. For example, in the DCT case, a block of 8 × 8 pixels is converted into an 8 × 8 array of DCT coefficients.

The output of the typical signal analyzer process is usually quite accurate with 8 to 12 bit precision. In lossless compressers, this accuracy is preserved in the quantization process. In lossy systems, the quantizer reduces the accuracy of the transformer output in a way that is as acceptable to the eye as possible. The quantization can be employed on either a scalar or vector basis. Most of the compression is achieved in this step by coarsely quantizing the transformed signal in such a way that any distortion to the eye is minimized. Typically, large numbers of coefficients are discarded because their value is low.

The final step in the compression process is to encode the quantizer output with a variable length code (VLC, sometimes called entropy coding). VLC is a technique whereby each event is assigned a code that may have a different number of bits. To obtain compression, short codes are assigned to frequently occurring events and long codes are assigned to infrequent events. Table 3.3 illustrates a variable length code for a set of four symbols. When symbol A is transmitted half the time, a compression of 2 to 1 is achieved. When symbols C or D are transmitted, a bit rate expansion of 50% is incurred. The expectation is that the average code length will be less than the fixed code that would otherwise be required. A major advantage of VLC is that it does not degrade the signal quality in any way (lossless). Therefore, the output picture quality is transparent to the VLC used.

Table 3.3
Huffman Variable Length Coding

Symbol	Probability	VLC Code	Fixed Length Code
A	1/2	0	00
B	1/4	10	01
C	1/8	110	10
D	1/8	111	11

Table 3.4 is a summary of the various levels of video compression which are employed for different applications. Level 1 is the uncompressed PCM level that is used as a reference to measure the compression for all other levels. The bit rate for this reference is approximately 90 Mbps which is derived from the fact that an NTSC video signal must be sampled at a rate over 11×10^6 samples/s and coded with 8 bits/sample. Level 2 is a true mathematically lossless compression system requiring a bit rate of approximately 45 Mbps. There are very few applications for such a system where the original PCM signal is preserved exactly. Level 3 is visually lossless, providing no visual degradation, and essentially providing broadcast TV quality. Level 4 is characterized by a 2-to-1 resolution reduction and an occasional perception of picture degradation when rapid detailed motion is encountered. Level 5, which is typically used for project and engineering management videoconferences, is characterized by more frequent occurrence of image degradation, but it is considered quite acceptable for these applications. In level 6 the resolution is further reduced.

3.4 Predictive Coding

PCM transmits each pixel as an independent sample without taking advantage of the high degree of pixel-to-pixel correlation existing in most pictures. Predictive coding is a basic bit rate reduction technique that does reduce this pixel-to-pixel redundancy. Figure 3.8 is a block diagram illustrating the basic predictive coding process. A predictor predicts the brightness value of each new pixel based solely upon the pels that have been previously quantized and transmitted. The predictor can vary greatly in complexity. The new pixel can be predicted from the previous pixel in the line, or a block of pixels can be predicted from a block of pixels in the previous video frames. The predicted brightness

Table 3.4
Levels of Video Compression

Compression Level	Bit Rate	Compression Ratio
Uncompressed	90 Mbps	1
True lossless compression	45 Mbps	2
Visually lossless	5–10 Mbps	10–20
High quality videoconferencing	384 Kbps to 1.5 Mbps	50–200
Acceptable quality videoconferencing	100 Kbps	1,000
Videophone	25 Kbps	4,000

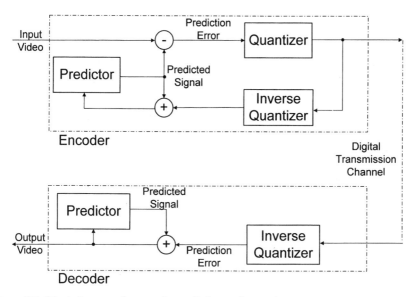

Figure 3.8 Block diagram of a generic predictive coding system.

value is subtracted from the actual brightness value of the new pel resulting in a bipolar prediction error signal. This error signal is quantized and transmitted. This quantization process can vary over a wide range of complexities. The most common technique is conventional pel-by-pel scalar quantization. However, it is also possible to employ transform coding or vector quantization. The quantization can be fixed, or it can adapt to the data itself. The quantizer can also vary over a wide range of accuracies. If 1-bit quantization is employed, the system becomes the well-known Delta Modulation technique. If the predictive quantizer employs multiple bits per pel, the technique is commonly defined as differential PCM (DPCM). At the receiver the inverse of the quantization process is performed and the decoded error signal is added to the predicted value to form the output signal for viewing. The output signal is fed to the predictor to be used for prediction of the next pel. Referring back to the predictive encoder, the reader will note that the transmitted signal is decoded at the transmitter using exactly the same decoding process that is used at the receiver. The predictive encoder can be viewed as a servo loop that continually forces the decoded output signal to be as close as possible to the input signal.

Figure 3.9 illustrates the transfer function of a typical 3-bit DPCM predictive coder. The quantizer is usually nonlinear because the eye is very sensitive to small changes in low-detail portions of a picture (small prediction error), but the eye is insensitive to coarse quantization of high-contrast edges (large predictive error). The design of this quantizer is a compromise between conflicting

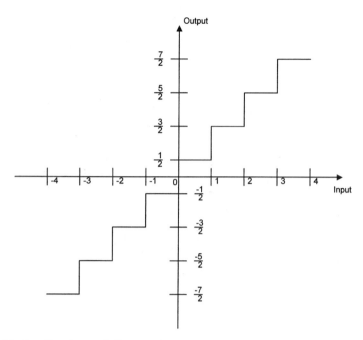

Figure 3.9 Quantizer for predictive encoder.

objectives. It is desirable that the quantizer precision be fine, particularly for small error signals, to keep the background granular noise in the output picture at an acceptably low level. On the other hand, the quantizer steps must be large enough, particularly the largest increment, so that the output can respond reasonably well to high-contrast changes in the input picture. If the largest increment is too small, slope overload occurs, resulting in picture blurring.

3.5 Discrete Cosine Transform

Transform coding algorithms, generally speaking, operate as two-step processes. In the first step, a linear transformation of the original signal (separated into subblocks of $N \times N$ pels each) is performed, in which pixel space is mapped into transform space. In the second step, the transformed signal is compressed by quantizing each subblock of transform coefficients. The reconstruction operation involves performing an inverse transformation of each decoded transformed subblock. The function of the transformation operation is to make the transformed samples more independent than the original samples so that the subsequent operation of quantization may be done more efficiently.

The transformation operation itself does not provide compression; rather, it is a remapping of the signal into another domain in which compression can be achieved more effectively. Compression can be achieved for two reasons. First, not all of the transform domain coefficients need to be transmitted to achieve acceptable picture quantity. Second, most of the coefficients that are transmitted can be encoded with reduced precision without seriously affecting image quality.

3.5.1 Transformation Techniques

Transforms that have proven useful include the Karhunen-Loeve transform (KLT), discrete Fourier transform (DFT), DCT, and Walsh-Hadamard transforms. The KLT is considered to be an optimum transformation, and for this reason many other transformations have been compared to it in terms of performance. However, the KLT has certain characteristics that make it less than ideal for image processing. For data having high interelement correlation, the performance of other transforms (such as the DCT) is virtually indistinguishable from that of the KLT and, thus, usually does not warrant its added complexity.

The DFT is one of the few complex transforms used in data coding schemes. There are disadvantages to using a complex transform for data coding, the most obvious of which is the storage and manipulation of complex numbers. Again, as in the case of the KLT, this complexity issue would not be a factor if the performance of the DFT were significantly greater than that of other transforms. However, other transforms that are less complex perform better than the DFT.

The DCT is one of an extensive family of sinusoidal transforms. In their discrete form, the basis vectors consist of sampled values of sinusoidal or cosinusoidal functions that, unlike those of the DFT, are real number quantities. The DCT has been singled out for special attention by workers in the image processing field, principally because for conventional image data having reasonably high interelement correlation, the DCT's performance is virtually indistinguishable from that of other transforms that are much more complex to implement. The three transforms mentioned previously have basis functions that are either cosinusoidal (that is, the Fourier and discrete cosine) or are a good approximation of a sinusoidal function, such as the KLT. The Walsh-Hadamard transform is an approximation of a rectangular orthonomal function. The actual transform consists of a matrix of $+1$ and -1 values, which eliminates multiplications from the transform process. The elimination of multiplications is a significant property, since the aforementioned transforms require real or complex multiplications. However, the Walsh-Hadamard transform does not provide the excellent performance that the DCT provides.

Since the DCT is universally accepted as the preferred transform for image coding, it is useful to provide more detail on its implementation. The execution of the DCT algorithm requires the division of an image into a series of (N × N) subblocks of pixels. Each subblock is transformed by a two-dimensional (N × N) DCT process as (T) = (C) + (D) = (C)^T where (T) is the transformed subblock, (C) is the DCT basis matrix, and (D) is the input data subblock ((C)^T is the transpose of the DCT basis matrix). The DCT basis matrix coefficients were determined from the relation

$$C_{1 \cdot j} = C_o \cdot \left(\frac{2}{N}\right) \cdot \left\{\cos\left[I \cdot (j + 0.5) \cdot \left(\frac{n}{N}\right)\right]\right\}$$

where $C_o = \frac{1}{2}$ for $1 = 0$, $C_o = 1$ otherwise, and $I = j = 0$ to $N - 1$. Figure 3.10 illustrates the basis functions for a 8 × 8 DCT. This transformation converts each (N × N) subblock of pixels into an (N × N) matrix of transform

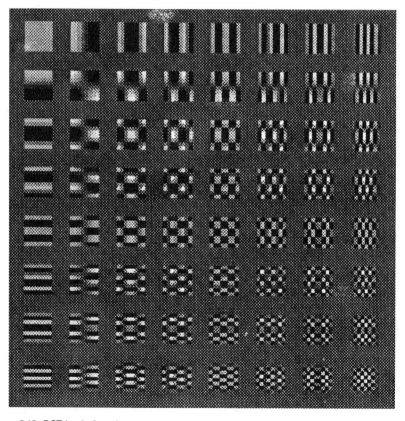

Figure 3.10 DCT basis function.

coefficients, which consists of one DC coefficient and (N × N − 1) AC coefficients. The sum of the squares of all of the AC coefficients in a given transform matrix is known as the AC energy of that transform matrix.

3.5.2 Coding of DCT Coefficients

As explained in the previous section, the DCT is usually used when pictures are transmitted using transform techniques. This transformation merely creates a set of coefficients equal in number to the original set of pels. At this point no compression has been accomplished except that the original set of pels with uniformly high redundancy have been decorrelated and the information has been compacted in the lower spatial frequency coefficients. The purpose of this section is to address the second part of the two step process, that is, how to encode the transform coefficients for transmission.

The first step in the coding process is to determine which coefficients are to be transmitted and which are to be deleted. Figure 3.10 illustrates the set of 64 transform coefficients corresponding to a 8 × 8 block of pels to be coded. Coefficient number one is the DC coefficient that is a measure of the average brightness of the block. Coefficients in the top row measure of spatial frequency content in the horizontal direction. Coefficients in the left column measure frequencies in the vertical direction, and all others measure various combinations thereof. In general, most of the energy is contained in the low-frequency coefficients with relatively little signal strength in the high-frequency coefficients.

Figure 3.11 shows a simple example of how each 8 × 8 block is coded. Figure 3.11(a) shows the original block to be coded. The block has a constant slope or shading from the upper left-hand corner to the lower right. Without compression, this would take 8 bits to code each of the 64 pixels or a total of 512 bits. First, the block is transformed, using the two-dimensional DCT, giving the coefficients of Figure 3.11(b). Note that most of the energy is concentrated into the upper left-hand corner of the coefficient matrix.

Essentially, the DCT is performed by multiplying the input block by each of the 64 basis functions shown graphically in Figure 3.10. The results of each of these multiplications, also 8 × 8 arrays, are summed to give the 64 transform coefficients. In the upper left-hand corner of Figure 3.10, the first basis function is constant over the block and therefore gives rise to the dc value of the input block. At the opposite corner, the basis function is a checkerboard and will give significant coefficient values only if there are elements of this pattern in the input block. Of course, the coefficients are in practice calculated by a chip in a more efficient manner than described here.

Next, the coefficients of Figure 3.11(b) are quantized with a stepsize of 6. (The first term (DC) always uses a stepsize of 8.) This produces the values of

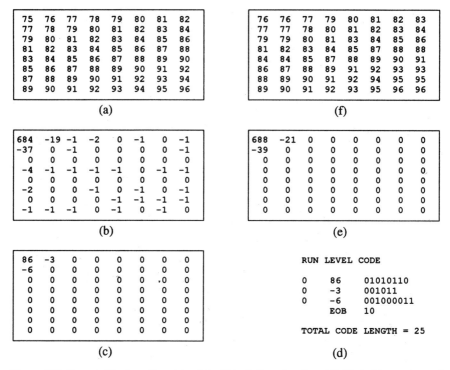

(a)

(f)

(b)

(e)

(c)

RUN LEVEL CODE

RUN	LEVEL	CODE
0	86	01010110
0	-3	001011
0	-6	001000011
	EOB	10

TOTAL CODE LENGTH = 25

(d)

Figure 3.11 Sample block coding; (a) original block (8 × 8 × 8 = 512 bits); (b) transformed block coefficients; (c) quantized coefficient levels; (d) coefficients in zig-zag order and variable length code; (e) inverse quantized coefficients; and (f) reconstituted block.

Figure 3.11(c), which are much smaller in magnitude than the original coefficients and most of the coefficients become zero. The larger the stepsize, the smaller the values produced, resulting in more compression.

The coefficients are then reordered, using the zigzag scanning order of Figure 3.12. All zero coefficients are replaced with a count of the number of zeros before each nonzero coefficient (RUN). Each combination of RUN and VALUE produces a VLC that is sent to the decoder. The last nonzero VALUE is followed by an end of block (EoB) code. The total number of bits used to describe the block is 25, a compression of 20:1. The code in Figure 3.11(d) is based upon a two-dimensional VLC (ITU-T Recommendation H.261) shown in Table 3.5.

Additional examples in Table 3.6 illustrate the coding process. In examples 1, 2, and 3, the codes are obtained directly from Table 3.5. For examples 4, 5, 6, and 7, the combinations of run length and level cannot be found in Table 3.5, and so the escape code is used, together with the binary representation of the run length (6 bits) and the 2s complement representation of the level.

1	2	6	7	15	16	28	29
3	5	8	14	17	27	30	43
4	9	13	18	26	31	42	44
10	12	19	25	32	41	45	54
11	20	24	33	40	46	53	55
21	23	34	39	47	52	56	61
22	35	38	48	51	57	60	62
36	37	49	50	58	59	63	64

Figure 3.12 Scanning order in a block.

At the decoder, the stepsize and VALUEs are used to reconstruct the inverse quantized coefficients, which, as shown in Figure 3.11(e), are similar to but not exactly equal to the original coefficients. When these coefficients are inverse transformed, the result of Figure 3.11(f) is obtained. Note that the differences between this block and the original block are quite small.

Figure 3.13 shows a slightly different example illustrating more clearly some of the features of DCT coding. In this example, in addition to shading, the block contains a checkerboard pattern that matches the highest order basis function. This causes the last coefficient to be transmitted. There are 60 zero-value coefficients (in the zigzag order) between the previous nonzero coefficient and the last one, so the run length is 60. The last coefficient is coded as a 6-bit escape code (000001), a 6-bit run code (111100) [6010], and an 8-bit level code (00000010) [210] as shown at the bottom of Table 3.4.

Two types of distortion appear in transform-coded pictures: truncation errors and quantization errors. Quantization errors are noiselike, whereas truncation errors cause blocking distortion and a loss of resolution. In practice, the truncation threshold and quantization precision must be adjusted experimentally to achieve the maximum compression and acceptable picture quality. In

Table 3.5
Two-Dimensional VLC Table

Run	Level	Code
EoB	1	10
0	1	1sa If first coefficient in block
0	1	11s Not first coefficient in block
0	2	0100 s
0	3	0010 ls
0	4	000 110s
0	5	0010 0110 s
0	6	0010 0001 s
0	8	0000 0001 1101 s
0	9	0000 0001 1000 s
0	10	0000 0001 0011 s
0	11	0000 0001 0000 s
0	12	0000 0000 1101 0s
0	13	0000 0000 1100 1s
0	14	0000 0000 1100 0s
0	15	0000 0000 1011 1s
1	1	011s 1 2 0001 10s
1	3	0010 0101 s
1	4	0000 0011 00s
1	5	0000 0001 1011 s
1	6	0000 0000 1011 0s
1	7	0000 0000 1010 1s
2	1	0101s
2	2	0000 100s
2	3	0000 0010 11s
2	4	0000 0001 0100 s
2	5	0000 0000 1010 0s
3	1	0011 1s
3	2	0010 0100 s
3	3	0000 0001 1100 s
3	4	0000 0000 1001 1s
4	1	0011 0s
4	2	0000 0011 11s

Table 3.5 *(continued)*

Run	Level	Code
4	3	0000 0001 0010 s
5	1	0001 11s
5	2	0000 0010 01s
5	3	0000 0000 1001 0s
6	1	0001 01s
6	2	0000 0001 1110 s
7	1	0001 00s
7	2	0000 0001 0101 s
8	1	0000 111s
8	2	0000 0001 0001 s
9	1	0000 101s
9	2	0000 0000 1000 1s
10	1	0010 0111 s
10	2	0000 0000 1000 0s
11	1	0010 0011 s
12	1	0010 0010 s
13	1	0010 0000 s
14	1	0000 0011 10s
15	1	000 0011 01s
16	1	0000 0010 00s
17	1	0000 0001 1111 s
18	1	0000 0001 1010 s
19	1	0000 0001 1001 s
20	1	0000 0001 0111 s
21	1	0000 0001 0110 s
22	1	0000 0000 1111 1s
23	1	0000 0000 1111 0s
24	1	0000 0000 1110 1s
25	1	0000 0000 1110 0s
26	1	0000 0000 1101 1s
Escape	1	0000 01

Table 3.6
VLC Coding Examples

Example	Run Length	Level	Code
1	0	3	001010
2	0	−3	001011
3	26	−1	0000 0000 1101 11
4	32	−1	000001,100000,11111111
5	60	2	000001,111100,00000010
6	60	−2	000001,111100,11111110
7	60	−86	000001,111100,10101010

general, transform coding is preferable to predictive coding for compression to bit rates below 1 or 2 bits-per-pel for single pictures. However, in those applications where cost and complexity are important issues the choice between these two algorithms may be less clear.

3.5.3   DCT Interframe Prediction With Motion Compensation

In 1990, the ITU approved the H.261 video coding standard which is the first standard based on DCT Interframe Prediction with Motion Compensation. This basic coding algorithm has been subsequently employed for virtually every

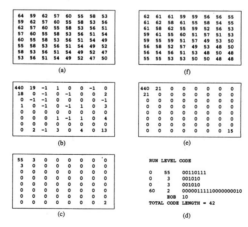

Figure 3.13 Example of DCT coding: (a) original block; (b) transformed block coefficients; (c) quantized coefficient levels; (d) coefficients in zig-zag order and variable length code; (e) inverse quantized coefficients; (f) reconstituted block.

video coding standard including MPEG-1, MPEG-2, the U.S. HDTV standard, satellite broadcast TV, and DVD. The coding algorithm basically develops a prediction for an 8×8 block of pixels based upon previously transmitted blocks. The DCT coefficients of the block to be transmitted are subtracted from the DCT coefficients of the prediction block, and the residual prediction error is transmitted. If the prediction is accurate, the residual error is small, no information needs to be transmitted, and the compression is extremely high. If a significant residual error exists, it is coded, transmitted to the receiver where it is added to the predicted block coefficients, decoded, and the block is displayed.

This coding algorithm, known as DCT Interframe Prediction, works very well if there is little or no motion is the scene. Motion would reduce the accuracy of prediction of those blocks where motion occurs, thereby reducing the compression. The problem of motion has been addressed by the addition of motion compensation to the algorithm. Motion compensation, illustrated in Figure 3.14, basically tracks the motion of moving objects in the input scene, and develops an interframe prediction, based on that block in the prediction picture which is most similar to the block being transmitted. A motion vector is transmitted to indicate the location of the block in the prediction image which is being used for prediction. Motion compensation is an extremely important element of this basic compression system contributing greatly to the picture quality achieved by the coder. Although motion compensation is mandatory at the decoder and optional at the encoder, it is employed in virtually every system in use today. Motion compensation is very simple to implement in the decoder, but is very complex to implement in the encoder due to the requirement to search for a best prediction match.

3.6 Vector Quantization

Quantization is the process by which an analog signal, having a continuous range of possible values, is divided into a limited set of discrete steps. The most common quantization procedure is scalar, where one can visualize a scale placed next to the variable being measured. In the scalar quantization process, one digital word represents the quantized value of one sample of a signal (for example, the value of a single pixel in an image).

In the case of vector quantization (VQ), one digital word represents the quantized value of more than one sample of a signal. In the case of the VQ of an image [1–9], a single word, or vector, is typically used to represent the quantized values of, for example, an entire 4×4 array of pixels. Unfortunately, it is difficult to visualize how one vector represents multiple quantized values in n-dimensional space. It is easier to visualize the vector quantization of an array

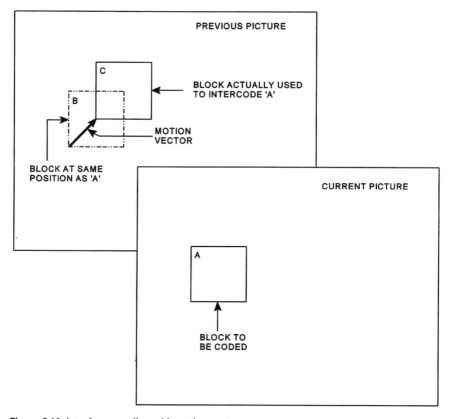

Figure 3.14 Interframe coding with motion vectors.

of two adjacent pixels as illustrated in Figure 3.15. The figure shows the case where each pixel is first quantized, by conventional scalar means, to 4-bit (16-level) precision. The figure illustrates one vector representing the combination of brightness value 6 for pixel A and brightness value 9 for pixel B. In general, the vector could take on 16 × 16 = 256 possible values; however, if only this procedure were employed, no compression would be achieved.

Compression in VQ systems is accomplished in two steps. In the first step, a number of adjacent vectors, which occur in a cluster in the input, are represented by a single vector at the center of the cluster. Clustering is accomplished using two approaches. In one case, a very large sample of actual vectors from real images, usually known as a training vector set, is accumulated. From this set, vector clusters are located by rather complex algorithms, and representative vectors are placed at the centers of these clusters. The trade-off between distortion and compression is dependent on how many representative vectors

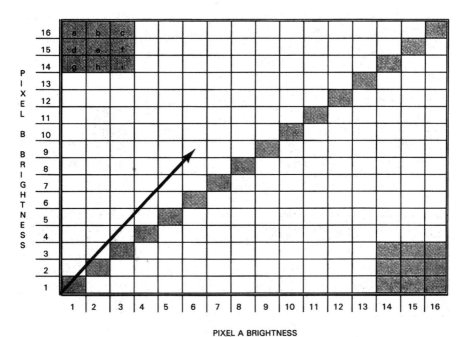

PIXEL A BRIGHTNESS

Figure 3.15 Vector quantization.

are included in the list. The longer the list, the less the distortion, but also the less the compression.

In the second approach, clusters are defined on the basis of psychovisual perception. For example, the eye is very insensitive to variations in high-contrast edges such as those occurring in the cluster of nine vectors (a through i) in the corner of Figure 3.16. If any of these nine vectors (a through i) occur in the input image, one vector representing the cluster of vectors could be transmitted, and brightness values corresponding to vector e would be displayed at the output, without any perceived distortion. Little, if any, clustering is accomplished for equal brightness vectors (45 degree diagonal in Figure 3.16) because they occur frequently and the eye is sensitive to distortions in this region.

The lossy compression previously described merely sets the stage for the second step of compression, which is lossless VLC (also known as entropy coding). This coding takes advantage of the fact that the vectors to be transmitted are not equally likely. Short codes are assigned to the most likely vectors, and long codes are assigned to the least likely. In this way the overall VQ compression is a combination of psychovisual lossy compression and mathematical lossless compression. The above example illustrates VQ compression for a single image in two dimensions, that is, x, y. Clearly VQ can be used even more effectively to compress a video signal in three-dimensions, that is, x, y, t.

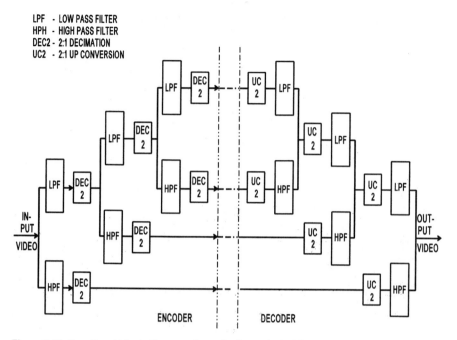

Figure 3.16 Functional block diagram of a typical wavelet/subband codec.

One important characteristic of VQ is that it is highly asymmetric. That is, the encoder is far more complex than the decoder. The encoder requires a complex search process to make the decision as to what vector to transmit, while the decoder is merely a look up table to display the value corresponding to the transmitted vector. This asymmetric characteristic makes VQ very attractive for applications where video is encoded once (for example, CD-ROM) and displayed many times on inexpensive display terminals.

3.7 Wavelet and Subband Coding

Subband coding and wavelet coding are closely related and operate by decomposing an image into frequency components [10–15]. This decomposition is performed by iterative low- and high-pass filtering. The result of the high-pass filter is retained. The result of the low-pass filter is decimated and again low- and high-pass filtered. This is continued for the number of levels desired (typically 3 or 4). See Figure 3.16.

This filtering is performed on the entire image, not on small blocks like the DCT. This eliminates the blocking artifact normally associated with DCT algorithms. At each frequency band, there is a filtered version of the image whose spatial resolution is proportional to the frequency. For example, low-frequency bands have low spatial resolution that represent larger areas of the image. High-frequency bands have high spatial resolution that represent local edge and texture detail. While the DCT coefficients represent a decomposition based on frequency-related cosines, the wavelet transform is a decomposition based on frequency-related complex waveforms or wavelets. The result of the wavelet transform is a hierarchical representation of the image where each level in the hierarchy is a map of the image information contained within a specific frequency band. This hierarchy is represented by a tree structure of filtered pixels, with the lowest frequency pixels at the root. Each pixel at one level is linked to multiple pixels at the next level. The lowest frequency level is the called the reference signal. The other levels are the detail signals.

There are two important steps in designing a wavelet compressor: selection of the wavelet filter and effectively quantizing the resulting tree structure. The wavelet filter is divided into two parts, an analysis filter and a synthesis filter. The analysis filter is used during compression, and the synthesis filter is used during decompression. This filter pair must meet certain constraints. These are perfect reconstruction, finite length, and regularity. Perfect reconstruction indicates that a signal put into the input of the analysis filter is identical to the signal that comes out of the synthesis filter within some small error. Finite length means that it is implementable. Regularity means that the filter converges to a continuous function. Meeting these criteria, the filter is then selected to pack as much information about the original image as possible into the reference signal. This is because the reference signal is generally quantized finer than the detail signals, which are quantized coarsely or discarded.

As with the DCT, the quantizer is the function where the compression gain is realized. The transform converts the pixels into a format that is more compressible. The quantizer does the compression. After quantizing the reference signal and detail signals, one is left with a sparsely populated tree of spatial and frequency position-dependent information. This information is further compressed by effectively representing only the significant data without losing the position information. This is done by tree trimming algorithms or run length coding. This leads to large compression ratios while maintaining image quality and detail.

In decompression, the inverse quantized, filtered reference signal, and detail signals are up sampled (interpolated) and applied to the synthesis filter banks. The outputs of these banks are summed. See Figure 3.16. The result is the original image less any distortion due to quantization.

In wavelet compression, the artifacts that occur at high compression ratios are loss of detail and ringing or oscillation around edges. The loss of detail is a result of the coarse quantization of the detail signals. The ringing is a result of the characteristics of the wavelet filters. Those filters exhibiting ringing in the impulse response of the filter will exhibit ringing to a greater extent in the resulting image.

There is work currently being done to adapt wavelet compression to motion video. There is no single approach being used for interframe compression using wavelets. Several organizations, both commercial and academic, are investigating different techniques for interframe coding. One such technique takes the wavelet transform of the previous image and subtracts it from the wavelet transform of the current image. It then quantizes and transmits only the differences.

Another technique for motion video compression using wavelets is region-based wavelet transforms. This technique identifies regions in the image. The regions are tracked for motion. Then wavelets are fitted to each region. Other techniques under study are adaptive wavelet transforms, overcomplete wavelet transforms, and zero crossing translates. All of these techniques attempt to account for motion components (for example, translations, rotations, and scaling) in the wavelet transform.

3.8 Fractal Coding

A fractal is a structure possessing similar looking forms of many different sizes and orientations. It can be magnified infinitely with structure at every scale and can be generated by small, finite sets of data and instructions. Fractal compression is based on three concepts: affine maps, iterated system functions, and the collage theorem. The affine map is a combination of rotations, scalings, and translations that act on a part of a source image to create a part of a target image. An iterated function system is a collection of contractive, affine maps. The iterated function system acts on a part of the source image to create a part of the target image out of repeated parts of the source image. These repeated parts are of various rotations, scales, and translations. The collage theorem says that if an image can be described by a set of affine maps then that set of maps provides an iterated system function that can be used to reproduce as good an approximation of the image as you desire.

The challenge is then to find a fractal model for a given image. This is done through the fractal transform, which is a systematic method that breaks up an image into smaller regions, called domain regions, and finds the best affine maps for those regions. The domain regions are nonoverlapping and completely cover the image. Range regions are also defined. They can overlap and do not need to cover the entire image. An affine map is generated that maps every range region into each domain region. The map and range region that provide the best match for a given domain region are selected. The affine coefficients, the domain region geometry, and the range region addresses for each domain range are packed to form the compressed data file, an FIF file, for still images or the compressed data stream for intracoded motion video. See Figure 3.17.

Decompression is performed by first creating two arbitrary buffers. The domain regions are identified on one buffer and the range regions on the other. The affine maps are applied using the range buffer and filling the domain buffer. The range and domain regions are then identified on the opposite buffers and the affine transforms are applied again. This is repeated until the differences between the two buffers are sufficiently small. This process results in an image that closely matches the original image before compression. See Figure 3.18.

Fractal coding of the difference images typically used in interframe coding does not work well because of the statistics of the difference image. One technique that is under study uses the difference image to identify areas where the prediction fails. The areas where prediction was satisfactory are defined as background; the areas where prediction failed are foreground. Those areas identified as background do not need to be transmitted since they are the same as the previous image within a tolerable error. The foreground areas are then fractal coded and transmitted.

3.9 Object-Based Coding Techniques

One difficulty with the H.261 coding algorithm is that blocking artifacts and mosquito noise are frequently generated at the boundary between moving objects and the static background. This is caused by the fact that the DCT block edges are not aligned with the edge of the object. Object-based coding (OBC) has been devised to attempt to minimize distortions of this type. The general principle of OBC [16–23] involves the identification of, and the encoding of, arbitrarily shaped objects moving within the scene. Information is transmitted about the object defining its motion, shape, and color. Two types of OBC systems have been investigated: a generic approach dealing with unknown objects, and approaches dealing with known objects such as a talking head (also known as knowledge-based coding). The status of work in these two areas is outlined as follows.

3.9.1 Generic Unknown Objects

The University of Hanover, in Germany, has been a leader in the research of OBC. Scholars there have published many articles on the subject and have focused most of their energies on a specific implementation known as object-based

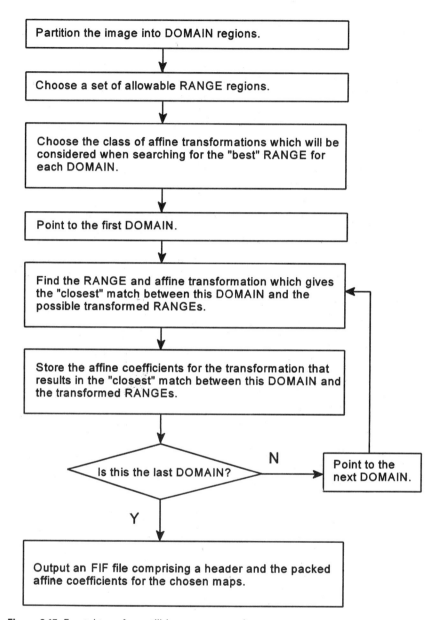

Figure 3.17 Fractal transform still-image compression.

analysis-synthesis coding (OBASC). Figure 3.19 is a functional block diagram of the OBASC encoder. The general architecture is the same as the H.261 coder in that they both employ interframe prediction and use the classic predictive loop structure. The parameter coder compares the current parameters with the stored

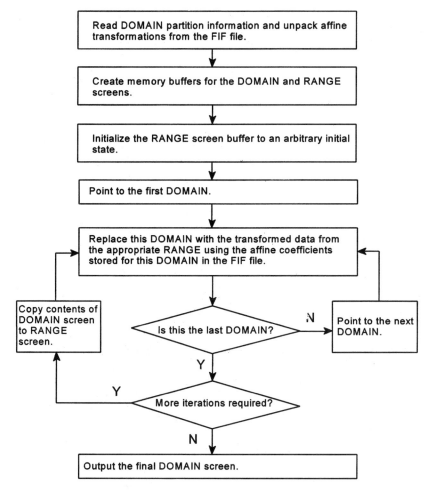

Figure 3.18 Fractal transform still-image decompression.

and predicted parameters and transmits the prediction errors to the receiver. The loop is a little more complex than normal because not only are current and predicted parameters being compared, but current and synthesized images are also being compared by the image analyzer. The image analyzer develops the current parameter set.

An example of image analysis for the test sequence "Miss America" is illustrated in Figure 3.20. The analyzer decomposes the input images of a sequence into differently moving objects. For each object, three sets of parameters are determined describing the object's shape, motion, and color. In addition, each object is classified as to whether it complies with the underlying source model (that

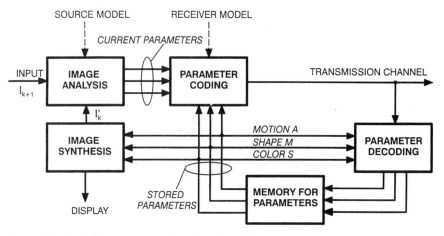

Figure 3.19 Block diagram of an object-based analysis-synthesis coder.

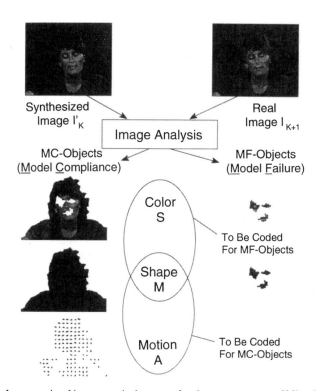

Figure 3.20 An example of image analysis output for the test sequence "Miss America."

is, whether the changes of the color parameters can be described only by that object motion that is allowed by the source model) or whether the source model fails. Figure 3.20 illustrates the processing of the two classes of objects—model compliant and model failure.

A great deal of research is being directed toward source models having a wide range of complexity. The simplest source model restricts the definition of the object to a rigid two-dimensional format. More advanced models are capable of defining the objects as having 2D/flexible, 3D/rigid, and 3D/flexible structures.

Research is also under way on a variation of OBC known as region-based coding. In general, a region is an area having uniform brightness. Consequently, an object is usually made up of a number of regions. The advantage of region-based coding is that the requirements for defining the texture of the region has been eliminated. The disadvantage is that the number of arbitrarily shaped elements to be defined and transmitted has increased.

3.9.2 Knowledge-Based Coding

The OBASC system previously described is designed to detect and encode any generic objects that may appear in the scene. It does not take advantage of any a priori knowledge; for example, it may be very likely that some objects (such as head and shoulders) will occur very frequently. Knowledge-based coding systems are specifically designed to take advantage of this prior knowledge. The knowledge-based object is usually defined by a wireframe model as illustrated in Figure 3.21. The first step in the encoding process is to detect the existence of the known object in the scene and the adaptation of the wireframe model to the characteristics of the particular object. At this point, the normal OBASC procedures of transmitting motion, shape, and color parameters take place.

Research is under way on a more advanced version of knowledge-based coding known as semantic coding. Semantic coding is applicable to a knowledge-based object that has a restricted set of action units. In the case of the head-and-shoulders object, a restricted set of action units could be mouth open/closed or eye open/closed. An example of semantic coding for the Miss America test scene is provided in Figure 3.22.

3.9.3 MPEG-4 (Coding of Audiovisual Objects)

Until recently, object coding has been a topic for research in the university rather than one for the development of standards. However, this has changed since the primary purpose of MPEG-4 is the development of standards for object coding. The MPEG-4 standard is being rolled out in two versions. Version 1, scheduled

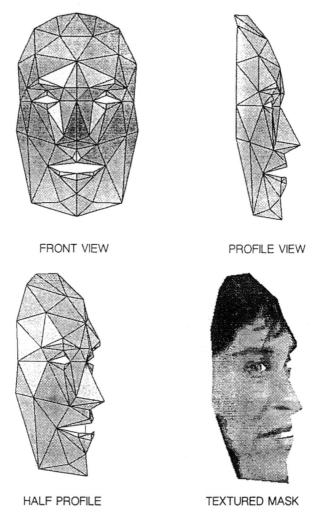

FRONT VIEW   PROFILE VIEW

HALF PROFILE   TEXTURED MASK

Figure 3.21 Model mask candide.

for approval in late 1998, will include face animation while Version 2, to be approved in late 1999, will provide body animation.

The face is an object capable of facial geometry ready for rendering and animation. The shape, texture, and expressions of the face are generally controlled by the bitstream containing instances of facial definition parameter (FDP) sets and/or facial animation parameter (FAP) sets. The structure of the FAP parameters is illustrated in Figure 3.23. Upon construction, the face object contains a generic face with a neutral expression. This face can already be rendered. It is also immediately capable of receiving the animation parameters

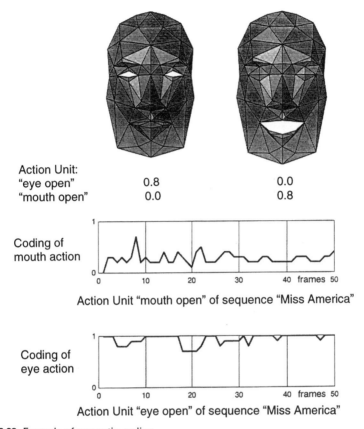

Action Unit: "eye open" 0.8 0.0
"mouth open" 0.0 0.8

Coding of mouth action

Action Unit "mouth open" of sequence "Miss America"

Coding of eye action

Action Unit "eye open" of sequence "Miss America"

Figure 3.22 Example of semantic coding.

from the bitstream, which will produce animation of the face: expressions, speech, and so on. If definition parameters are received, they are used to transform the generic face into a particular face determined by its shape and (optionally) texture. If so desired, a complete face model can be downloaded via the FDP set as a scene graph for insertion in the face node.

Face animation in MPEG-4 Version 1 provides for highly efficient coding of animation parameters whose decoding can drive an unlimited range of face models. The models themselves are not normative, although there are normative tools to describe the appearance of the model. Frame-based and temporal-DCT coding of a large collection of FAPs can be used for accurate speech articulation. Viseme and expression parameters are used to code specific speech configurations of the lips and the mood of the speaker. Depending on the application, facial animation can be used with or without the MPEG-4 systems layer.

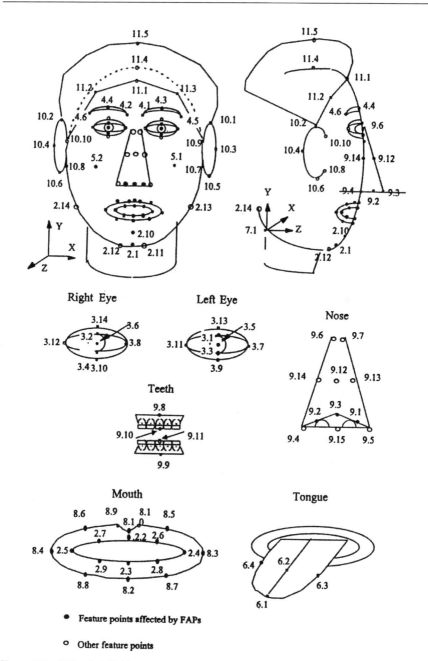

Figure 3.23 FAP unit definitions.

3.10 Variable Length Coding

VLC, also called entropy coding, is a technique whereby each event is assigned a code that may have a different number of bits. To obtain compression, short codes are assigned to frequently occurring events and long codes are assigned to infrequent events. The expectation is that the average code length will be less than the fixed code length that would otherwise be required. If all events are equally likely, or nearly so, then VLC will not provide compression.

All codes considered must be uniquely decodable; that is, there must be only one way that a concatenation of VLCs can be decoded. In addition, it is highly desirable that the code be instantaneous; that is, each code word can be decoded without reference to subsequent code words. Taken together, these requirements mean that no code word can be the beginning of another code word. For example, we may not have 01 and 0110 as code words, since the second code word starts with the first code word. In decoding, it is not known whether 01 is the first code word or just the start of the second code word.

A major advantage of VLC is that it does not degrade the signal quality in any way. The reconstituted signal will exactly match the input signal so that if the signal is adequately described by a series of events, using VLCs to communicate them to the decoder will not change the events. Therefore the system is transparent to the VLC used.

The disadvantage of VLCs is that they only provide compression in an average sense. Therefore, sometimes the code could be longer for a specific section of signal. This characteristic gives rise to the need for a buffer to match the variable rate of bit generation with the fixed bit rate of the communication channel and a control strategy to prevent long-term overflows or underflows of the buffer. Also the establishment of frames or packets of data becomes more difficult with VLCs.

Seven VLCs are discussed: comma codes, shift codes, B codes, Huffman codes, conditional codes, arithmetic codes, and two-dimensional codes.

3.10.1 Comma Code

The comma code is the simplest of the VLCs. It assigns to each event a different length of code, starting at 1. A particular bit polarity marks the end of the code word, such as in Table 3.7.

The advantage of the comma code is that it is simple to generate and decode, requiring only counters to count the number of ones. However, it is rare that this code accurately matches the statistics of the events, so it is used primarily where simplicity of implementation is important.

Table 3.7
Comma Code

Code
0
10
110
1110
11110
111110
.
.
.

3.10.2 Shift Code

In the case where the probabilities of the events decrease monotonically as the magnitudes increase, a great simplification can be obtained by using a systematic VLC, such as a shift code. In this code, each code word consists of a series of subwords, each of length L bits. The first subword is capable of conveying 2L values, one of which is a shift that indicates that the value of the code word is contained in the following subword. In this way, any length of code word can be obtained by concatenating a number of subwords together.

Table 3.8 contains examples of the beginnings of shift code tables for L = 1, 2, and 3, where a subword of all 1s indicates a shift. Note that for L = 1, the shift code reduces to the comma code. The shift code is best suited to cases where the probabilities drop off rapidly since the number of codes available only increases linearly with the length of the code. For example, for L = 3, there are seven codes with length 3 ($2^3 - 1 = 7$). Increasing the length to 6 only adds seven more codes.

3.10.3 B Code

Another variable length code that is systematic is the B code. Again the code consists of a sequence of subwords, each of length L. But in this case, one bit of the subword is used to designate whether another subword is to be added to the code word. Therefore the remaining L-1 bits in the subword can be used as part of the code. For L = 1, the B code also reduces to the comma code.

Table 3.9 contains examples of the beginning of B code tables for L = 1, 2, and 3.

Table 3.8
Shift Code

L = 1	L = 2	L = 3
0	00	000
10	01	001
110	10	010
1110	1100	011
11110	1101	100
111110	1110	101
1111110	111100	110
11111110	111101	111000
111111110	111110	111001
1111111110	11111100	111010
11111111110	11111101	111011
.	.	.
.	.	.
.	.	.
.	.	.

In this table, the asterick (*) marks the columns containing the continuation bits, where 1 indicates continue and 0 marks the last subword of the code word. This version of the B code is instantaneous. In another version, the continuation bit is the same value for all subwords in the code word but alternates with each succeeding code word. That version is not instantaneous.

The B code is best suited to cases where the probabilities drop off slowly since the number of codes available increases geometrically with increasing code length. For example, for L = 4, there are eight codes with length 4 (2L-1). Increasing the length to eight increases the number of codes by 64 ($2^{2(L-1)}$).

3.10.4 Huffman Code

The Huffman code is a VLC that provides the shortest average code length for a given distribution of input probabilities. The method for generating the code is well known, but a distribution of input probabilities, either theoretical or measured, is required before the code words can be calculated. If the actual distribution differs from that used to calculate the code, then the average code length may not be less than other codes. If a large enough sample can be obtained to measure the distribution accurately, the Huffman code may be an

Table 3.9
B Code

L = 1	L = 2	L = 3
**********	* * * *	* * *
0	00	000
10	01	001
110	1000	010
1110	1001	011
11110	1100	100000
111110	1101	100001
1111110	101000	100010
11111110	101001	100011
111111110	101100	101000
1111111110	101101	101001
11111111110	111000	101010
.	111001	101011
.	111100	110000
.	111101	110001
	10101000	110010
	10101001	110011
	10101100	111000
	10101101	111001
	10111000	111010
	10111001	111011
	10111100	100100000
	.	.
	.	.
	.	.

attractive choice. In any event, it provides a reference against which other codes can be compared, if the distribution is measured on the image being coded.

3.10.5 Conditional Variable Length Codes

In general, the most likely sample values to be encoded are near zero, and therefore the small values are given the shortest codes. However, zero is the most likely sample value only in the absence of information about other samples. If the values of neighboring samples are known, then the distribution of the current sample value can be markedly changed. In the simplest case, only the previous sample is used. Although, in principle, other samples can be used for each value of the previous sample, the frequency of occurrence of each of the current samples can be obtained, and a set of VLCs devised for each. Since both the encoder and decoder know the value of previous samples, decoding can take place without significant delay.

3.10.6 Arithmetic Coding

In arithmetic coding, the frequency of occurrence of the symbols to be coded is continuously measured by both the encoder and decoder. In the resulting code, there is not a one-to-one correspondence between the events and specific bits. It is possible to generate arithmetic codes at a rate of less than one bit per event, whereas a Huffman code requires at least one bit per event. Since arithmetic coders adapt dynamically to the statistics of the image being transmitted, the compression is generally superior to that for conventional nonadaptive VLCs.

3.10.7 Two-Dimensional VLC for Coding Transform Coefficients

A VLC has recently been developed that is particularly designed to code transform coefficients. It is a two-dimensional code where the two dimensions are the number of zero-value coefficients in a row (usually from a zig-zag scan) and the value of the next nonzero coefficient.

References

[1] "Component Vector Quantization," Annex 4 of CCITT Study Group VIII, Geneva, December 1–12, 1986.

[2] Abut, H., ed., "Vector Quantization," IEEE Acoustics, Speech, and Signal Processing Society, New York: IEEE Press, 1990.

[3] Gersho, A., "On the Structure of Vector Quantizers," *IEEE Trans. on Information Theory*, Vol. IT-28(2), Mar. 1982, pp. 157–166.

[4] Gersho, A., and B. Ramamurthi, "Image Coding Using Vector Quantization," *Proc. ICASSP*, Paris, 1982.

[5] Gersho, A., and R. M. Gray, *Vector Quantization and Signal Compression*, Norwell, MA: Kluwer Academic Publishers, 1992.

[6] Gray, R. M., and Y. Linde, "Vector Quantizers and Predictive Quantizers for Gauss-Markov Sources," *IEEE Trans. on Communications*, Vol. COM-30(2), Feb. 1982, pp. 381–389.

[7] Helden, J., and D. E. Boekee, "Vector Quantization Using a Generalized Tree Search Algorithm," *Proc. 5th Symp. Information Theory in the Benelux, ri Aalten,* May 1984, pp. 21–27.

[8] Huang, H. M., and J. W. Woods, "Predictive Vector Quantization of Images," *IEEE Trans. on Communications,* Vol. COM-33(11), Nov. 1985, pp. 1208–1219.

[9] Linde, Y., A. Buzo, and R. M. Gray, "An Algorithm for Vector Quantizer Design," *IEEE Trans. on Communications,* Vol. COM-28(1), Jan. 1980, pp. 84–95.

[10] Akansu, A. N., and R. A. Haddad, *Multiresolution Signal Decomposition: Transforms, Subbands, and Wavelets,* New York: Academic Press, Inc., 1992.

[11] Chui, C. K., "An Introduction to Wavelets," *Wavelet Analysis and Its Applications,* Vol. 1, New York: Academic Press, Inc., 1992.

[12] Zhang, Y.-Q., and S. Zafar, "Motion-Compensated Wavelet Transform Coding for Color Video Compression," *IEEE Trans. on Circuits and Systems for Video Technology,* Vol. 2, Sept. 1992, pp. 285–296.

[13] Ohta, M., and S. Nogaki, "Hybrid Picture Coding with Wavelet Transform and Overlapped Motion-Compensated Interframe Prediction Coding," *IEEE Trans. on Signal Processing,* Vol. 41, Dec. 1993, pp. 3416–3424.

[14] Hilton, M. L., B. D. Jaweerth, and A. Sengupta, "Compression Still and Moving Images With Wavelets," *IEEE Trans. on Image Processing,* Vol. 2, No. 3, 1994.

[15] Chui, C. K., ed., "Wavelets: A Tutorial in Theory and Applications," *Wavelet Analysis and Its Applications,* Vol. 2, New York: Academic Press, Inc., 1992.

[16] Hürtgen, B., and Th. Hain, "On the Convergence of Fractal Transforms," *Proc. IEEE Int. Conf. Acoustics Speech and Signal Processing ICASSP'94,* 1994.

[17] Hürtgen, B., F. Mller, and C. Stiller, "Adaptive Fractal Coding of Still Pictures," *Proc. Int. Picture Coding Symp.* PCS'93, Lausanne, Switzerland, 1993.

[18] Jacquin, A. E., "A Novel Fractal Based Block-Coding Technique for Digital Images," *Proc. IEEE Int. Conf. Acoustics Speech and Signal Processing ICASSP'90,* Vol. 4, 1990, pp. 2225–2228.

[19] Jacquin, A. E., "Image Coding Based on a Fractal Theory of Iterated Contractive Image Transformations," *IEEE Trans. on Image Processing,* Vol. 1, No. 1, Jan. 1992, pp. 18–30.

[20] Lundheim, L., *Fractal Signal Modelling for Source Coding,* Ph.D. thesis, Universitetet I Trondheim Norges Tekniske Høgskole, 1992.

[21] Aizawa, K., H. Harashima, and T. Saito, "Model-Based Analysis Synthesis Image Coding (MBASIC) System for a Person's Face," *Signal Processing: Image Communication,* Vol. 1, No. 2, Oct. 1989, pp. 139–152.

[22] Hötter, M., and R. Thoma, "Image Segmentation Based on Object Oriented Mapping Parameter Estimation," *Signal Processing,* Vol. 15, No. 3, Oct. 1988, pp. 315–334.

[23] Hötter, M., "Optimization and Efficiency of an Object-Oriented Analysis-Synthesis Coder," *IEEE Trans. on Circuits and Systems for Video Technology,* Vol. 4, No. 2, Apr. 1994, pp. 181–194.

4

Audio Coding

One important component of any audiovisual terminal for videoconferencing, videophone, and general multimedia applications is the speech coder. Indeed, it is generally agreed that the audio component is more critical than the video portion of the transmission. This chapter has three parts: (1) the attributes of speech coding, (2) a review of currently available speech coders from the ITU, (3) and a brief overview of MPEG audio coders. It should be noted that echo cancellation, although not standardized, is a very important part of any audio conferencing system. The basics of echo cancellation are treated in Chapter 2.

4.1 Speech Coder Attributes

Speech coders have four attributes: bit rate, quality, complexity, and delay. For a given application, some of these attributes are predetermined while trade-offs can be made among the others. For example, quality can usually be improved by increasing bit rate or complexity and sometimes by increasing delay. In the following subsections, the various attributes, with particular relevance to low bit-rate speech coding, are discussed.

4.1.1 Bit Rate

Public-switched telephone network (PSTN) video telephones are expected to operate at bit rates up to at least 33.6 Kbps. The higher the bit rate, the better the video quality. Given this fact, it is desirable that the speech coder use as little of the total bit rate as possible. Prior to Recommendations G.729 and G.723.1, ITU-T speech coding recommendations only existed for bit rates of 16 Kbps and higher. Rates lower than 9.6 Kbps have been used for digital cellular telephones, secure telephones, and satellite telephone services, such as Inmarsat provides.

4.1.2 Delay

Low-rate speech coders can be considered block coders. That is, they encode a block of speech, also known as a frame, at a time. Depending on the application, the total speech coding delay of the system is some multiple of the frame size. The minimum delay of the system is usually 3 or 4 times the frame size. In the case of low bit-rate speech coders, the total delays are far larger than those coders currently standardized in ITU-T G series recommendations. For example, many lower rate coders have frame sizes of 20 ms, resulting in a one-way delay of 60 to 80 ms. By contrast, 16-Kbps G.728 has a one-way delay of under 2 ms.

Even higher delays may result from the internal architecture of the overall videophone system. The delay of the video is usually much larger than that of the audio. For example, a video coder running at 5 frames per second has a frame size of 200 ms. The same 3 or 4 times the frame size rule also applies for the video coder. Thus, the minimum delay of the video is 600 to 800 ms in this case. The speech coder would have much less delay than the video coder. From experience with satellite links, we know that if the one-way speech coder delay is greater than 300 ms, users will notice and some will object. This is true even if there are no echoes. The echo problem is greatly exacerbated by large delays. Based on this evidence, all of the speech coders considered herein have reasonable delays. In fact, they might even be able to double their delay without having a perceived degradation in their quality. However, echo cancellers will be necessary because of the delay. As delay is increased, the inserted echo path loss requirements increase significantly. Therefore, there is a penalty in system design as delay is increased.

4.1.3 Complexity

Most speech coders are implemented first on digital signal processing (DSP) chips and then may later be implemented on special purpose VLSI devices. Speed and random access memory (RAM) are the two most important contributors to complexity. Generally one word of RAM takes up about as much space as six words of read only memory (ROM). In rating the complexity, the utilized scale is a combination of the following parameters of fixed-point DSP chips: (1) execution speed (million instructions per second (MIPS)), (2) RAM, (3) ROM.

DSP chips also come in floating point. Speeds are typically slower, but more can be executed in a single instruction cycle. They usually require off-chip RAM, and the combination of external RAM and larger size results in larger power usage and greater overall system cost.

4.1.4 Quality

The very first digital speech coding standard is Recommendation G.711 for 64-Kbps PCM speech. The distortion introduced by a G.711 codec is considered one quantization distortion unit (QDU). A speech quality experts group (SQEG) uses the QDU for network planning purposes. The second digital speech coding standard was G.721 32-Kbps adaptive differential pulse code modulation (ADPCM). This coder must be used in combination with a G.711 codec at its input and output. This combination is considered to have a distortion of 3.5 QDU. SQEG network planning guidelines call for a maximum of 14 QDU for an end-to-end international connection and less than 4 QDU for a domestic connection. The G.728 16-Kbps LD-CELP coder is considered to have the same QDU as G.721. Prior to G.723.1 and G.729, these are the only coders considered toll quality.

Regional bodies have created standards for digital cellular telephony. The first generation of digital cellular standards all produce clear-channel speech quality that is roughly comparable. Listeners can tell that the speech is not as clear as the original 64-Kbps coder, or as clear as G.721 or G.728. However, this level of quality is considered acceptable for cellular service and most users can hold a conversation for an extended period of time without listener fatigue due to the additional impairments. In 1995, several of the first generation cellular standards were upgraded to give better speech quality.

Lower bit-rate speech coders have been created for other applications. In 1990, Inmarsat held a contest to determine a combined speech and channel coder for mobile satellite. The combined rate was 6.4 Kbps, but only about 4.15 Kbps was used for the speech coder. In 1989, work was done to create a U.S. federal standard for secure telephones operating at 4.8 Kbps. In both of these cases, the quality of the speech was less than IS-54 VSELP or RPE-LTP. The speech is still quite intelligible, but there is enough additional distortion that listeners have to strain to understand some words. This strain causes listener fatigue during a prolonged conversation. Previous ITU-T speech coders and those used in cellular standards have primarily been tuned for telephone handset use. In a videophone, they may also be used with a speakerphone. Low bit-rate speech coders can be tuned according to how they are used. A setting that sounds best for a handset may not sound that way for a loudspeaker. In addition, the microphone characteristics of a handset and a speakerphone are different. This means the input speech spectra for the two are different. Since a videophone coder could be used both ways, any coder selected should be tested for both modes and tuned to give the best overall compromise. In addition to clear channel quality, there is the issue of how a speech coder behaves on a channel having bit errors. In the case of digital cellular standards, provision is made for additional bits to be used for channel coding to protect the information bearing bits.

Not all bits of a speech coder are equally sensitive. It is common practice to have two or three classes of bits with the most sensitive bits getting the most protection while the least sensitive class receives no protection at all. For low bit-rate applications such as secure telephones over 4.8-Kbps modems, it is reasonable to expect that the distribution of bit errors would be random. For high-rate modem signals, such as those used for video telephony, errors are more likely to occur in bursts due to the more complex modulation schemes used in these modems. In this case, the speech coder requires a mechanism to recover from an entire lost frame. This is referred to as a frame erasure concealment.

For the purpose of providing better video quality many videophones should use voice activity detectors with their speech coders. During nonspeech intervals the speech coder is turned off and the bits are allocated to the video coder to provide better video quality. At the receiver, comfort noise is played out to simulate the background acoustic noise at the encoder. A similar method is used for some cellular systems and is also used in digital speech interpolation (DSI) systems. Most international phone calls carried on undersea cables or satellites use DSI systems.

There is some impact on quality when these techniques are used. Subjective testing can determine the degree of degradation. The methods typically suffer from two problems. If the voice activity detector is slow to detect speech, the onset of the speech sound is lost. This is referred to as front-end clipping. The performance of the voice activity detector is responsible for either the front-end clipping problem or its solution. The second problem occurs when there is an unusual acoustic noise in the background of the speech. The noise will interfere with the speech signal but will be eliminated during nonspeech intervals. This has a disconcerting effect on the listener and can impact intelligibility. Depending on the character of the background noise, unwarranted amounts of comfort noise may be inserted during the nonspeech intervals, resulting in the listener hearing more noise during nonspeech intervals.

4.1.5 Validation

Speech coding standards can be specified in different ways. At the very minimum, there are the bitstream standards. A good example of this is the LPC-10E vocoder used for secure voice terminals by NATO. At this point the standard is about 20 years old. Only the bitstream was ever specified. This meant that all of the parameters to be quantized were specified, together with their quantization tables and the order in which these parameters are to be transmitted. Over the years improvements have been made to both the encoder and the decoder to improve the quality and intelligibility of the output speech.

While this might seem to be a strength, it is also a weakness. It does not provide a customer with assurance that the hardware will have a specified level of quality. Over the years the governments purchasing this equipment have had to specify their own tests to determine if a vendor's equipment met their requirements. Since there were relatively few governments and contractors, this was still considered an adequate solution.

At the opposite end of the spectrum are bit exact specifications. Examples of such coders are G.721 (32-Kbps ADPCM) and the digital European cellular full rate standard (RPE-LTP). If digital speech samples are presented to a bit exact coder, then the coder will produce the same exact bitstream, regardless of the manufacturer of the coder. Bit exact standards have test vectors. A given input test vector must result in a given output test vector or a given bitstream. Any deviation will be considered a failure to conform to the specification. Consequently, the entire inner workings of the coder must be available to the implementer so that output test vectors and bitstreams will match those specified. This also helps assure both customers and equipment providers that the terminal hardware will achieve the specified level of performance.

There are other possibilities that lie between these two. For G.728 (floating point), a mathematically exact specification was given. If implemented in floating point, most implementations were able to use commercially available DSP chips. Their outputs for test vectors were within a certain deviation of the official test vectors. The deviations were far below those detectable by the human ear. All concerned were assured that the coder's performance fully met all requirements.

For the North American TDMA digital cellular standard IS-54, the description of the coder is not as precise. A set of test vectors exists, but no one has exactly matched the output test vectors for a fixed point implementation. However, if their implementation is close enough to the output test vectors, then a manufacturer can be fairly certain that the coder is correctly implemented. The validation procedure for IS-54 also contains the description of a subjective test that can be given to make certain that the coder performs as specified.

The important point to be made here is that validation procedures need to exist. A sanctioned method for establishing conformance with the specification assures the manufacturer that the product can be implemented solely on the basis of the published specification. It assures the customer that the performance will be within specification. Finally, it assures network service providers that their service will meet the specified quality goals.

4.2   ITU Standard Speech Coders (G.7XX)

The ITU has established six standards for the coding of speech signals which are summarized in Table 4.1. There are two different types of coding algorithms employed in these six recommendations: (1) waveform coding using sample-by-sample quantization, (2) linear prediction analysis-by-synthesis (LPAS) technology. Waveform coding technology is employed in standards G.711, G.721, and G.722 while the LPAS algorithm is used in standards G.728, G.729, and G.723.1. In the LPAS case, a linear prediction model of speech production is excited by an appropriate excitation signal in order to model the signal over time. The parameters of both the speech model and the excitation are estimated and updated at regular time intervals (e.g., every 20 ms) and used to control the speech model. Table 4.1 also describes two coders that are presently being actively investigated by the ITU, but the standardization process is not yet complete. Each of these currently available, and prospective, standards is described below.

4.2.1   G.711 (64 Kbps, PCM)

In 1977, the ITU took the first basic step in digitizing audiovisual information by establishing the G.711 Recommendation to encode toll quality speech (3-kHz bandwidth) at a bit rate of 64 Kbps. The standard specifies that the speech signal be sampled at the rate of 8 kilosamples per second and each sample be encoded with 8-bit precision, thus yielding a transmission bit rate of 64 Kbps. The standard specifies modes for coding the speech signal with both μ-law and A-law coding characteristics. The G.711 standard has the advantage of low complexity, robustness to channel errors, and excellent speech quality. It has been, and continues to be, the basic coding technique for videoconferencing systems operating at high bit rates, such at T1 transmission, where the high bit rate assigned to the audio component is not critical. The G.711 Recommendation is mandatory for virtually every H.3XX multimedia communication system to maximize system interoperability.

4.2.2   G.721 (32 Kbps, ADPCM)

In 1988, the ITU established the first standard for the compression of toll quality speech. G.721 specifies that the 3-kHz analog audio signal be encoded using

Table 4.1
ITU Audio Coding

Standard	Coding Algorithm	Application	Rates (Kbps)	Audio Band- width	Frame Size	Typical Delay[1]
Existing						
G.711 (1977)	PCM	GSTN telephony, ISDN telephony/videoconferencing	48, 56, 64	3 kHz	125 μs	$\ll$1 ms
G.721 (1988)	ADPCM	GSTN telephony	32	3 kHz	125 ms	
G.722 (1988)	Subband ADPCM	ISDN telephony, commentary audio	48, 56, 64	7 kHz	125	<2 ms
G.728 (1992)	LD-CELP	GSTN, ISDN telephony/videoconferencing	16	3 kHz	625 μs	<2 ms
G.729 (1995)	ACELP	GSTN telephony, wireless/PCS, videotelephony	8	3 kHz	10 ms	35 ms
G.723.1 (1995)	MPE/ACELP	GSTN videophone	5.3, 6.4	3 kHz	30 ms	97 ms
Planned						
G.WSC (1999)		ISDN telephony/videoconferencing, commentary audio, storage	16, 24, 32	7 kHz	$<$=20 ms	$<$=25 ms
4 Kbps codec (1999)		Wireless/PCS	4	3 kHz	$<$=20 ms	$<$=80 ms

[1]The sum of algorithmic and computational delay.

ADPCM for a transmission bit rate of 32 Kbps. The input to the G.721 encoder is PCM-coded audio as generated by the G.711 encoder. The standard provides for the coding of both A-law and μ-law speech signals.

Functional block diagrams of the G.721 encoder and decoder are included in Figures 4.1 and 4.2, respectively. Subsequent to the conversion of the A-law or μ-law PCM input signal to uniform PCM; a difference signal is obtained by subtracting an estimate of the input signal from the input signal itself. An adaptive 15-level quantizer is used to assign four binary digits to the value of the difference signal for transmission to the decoder. An inverse quantizer produces a quantized difference signal from these same four binary digits. The signal estimate is added to this quantized difference signal to produce the reconstructed version of the input signal. Both the reconstructed signal and the quantized difference signal are operated upon by an adaptive predictor which produces the estimate of the input signal, thereby completing the feedback loop. The speech quality and error sensitivity of G.721 are only slightly inferior to G.711.

4.2.3   G.722 (64-56-48 Kbps, Subband Coder)

The only present standard for 7-kHz speech coding is ITU Recommendation G.722. Its principal applications are teleconferences and video teleconferences. It is recommended for use with H.320 terminals on connections greater than 128 Kbps. The wider bandwidth (50–7,000 Hz) is more natural sounding and less fatiguing than telephone bandwidth (200–3,200 Hz). The wider bandwidth increases the intelligibility of the speech, especially for fricative sounds, like /f/ and /s/, which are difficult to distinguish for telephone bandwidth.

The G.722 coder is a two-band subband coder with ADPCM coding in both bands. It uses a 24 tap quadrature mirror filter to divide the signal into two bands as shown in Figure 4.3. Each is ADPCM coded and transmitted with an embedded 4-5-6 bit adaptive quantizer. Block diagrams of the lower- and upper-frequency subband encoders are included in Figure 4.4 and 4.5, respectively. The upper band uses an ADPCM coder with a 2-bit adaptive quantizer. The lower band uses an ADPCM coder. This makes bit rates of 48, 56, and 64 Kbps all possible. The 24 tap filter causes a delay of only 1.5 ms and is efficiently implemented using just 12 multiplications per decimated sample. With sample-by-sample ADPCM quantization in each band, 1.5 ms is also the delay of the coder. No voice activity detector is specified for this coder by the ITU.

The high fidelity of the lower band coding (up to 6 bits per sample) does a good job of masking the rather scratchy 2 bits per sample coding of the upper band. This example shows that some perceptual masking is possible between the telephone bandwidth and the higher frequencies. At the same time, it seems to indicate that the ear responds favorably to noticing the frequency content in

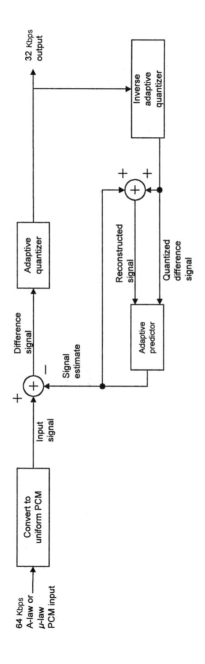

Figure 4.1 G.721 encoder.

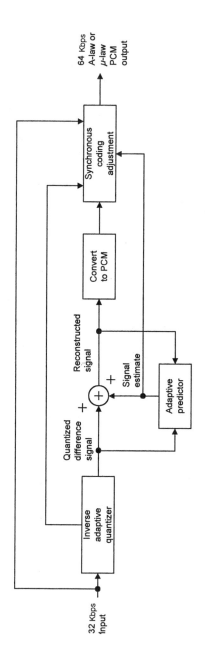

Figure 4.2 G.721 decoder.

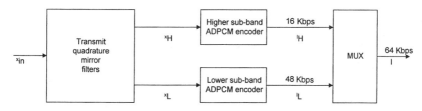

Figure 4.3 Block diagram of the G.722 encoder.

the upper band, even if it is scratchy. In subjective testing comparisons with telephone bandwidth speech using binaural listening (i.e., with both ears), the 8-kHz bandwidth is always preferred, sometimes by as much as 1 mean opinion score (MOS) point.

Recommendation G.722 is a low complexity coder. It was designed to be implemented on the first generation DSP chips. It requires about 3 MIPS and only a few hundred words of RAM. Since it was designed for the early fixed point DSP chips, it is specified in bit exact fixed point equations. Test vectors for G.722 are available from the ITU.

4.2.4   G.728 (16 Kbps, LD-CELP)

The ITU established the G.728 Recommendation in 1992 to provide toll quality speech at 16 Kbps. This standard has provided a major breakthrough for the videoconferencing application since the primary system bit rate for H.320 falls

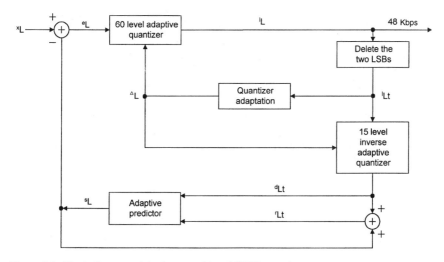

Figure 4.4 Block diagram of the lower subband G.722 encoder.

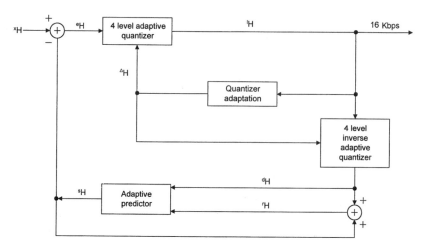

Figure 4.5 Block diagram of the higher subband G.722 encoder.

in the region of 112 to 384 Kbps. For systems operating at these low bit rates it is very important to minimize the audio bit rate to provide as much capacity as possible for video and data. This was accomplished when the G.728 standard was finalized.

The G.728 coding algorithm represents a sharp departure from the G.721 ADPCM coding technique. The algorithm is based on the codebook excited linear prediction (CELP) technique which is illustrated in Figures 4.6 and 4.7. Whereas ADPCM is a straight waveform-based coding technique, CELP is based somewhat on the characteristics of human speech and the properties of hearing. As with G.711 and G.721, the basic waveform sampling rate is 8 kilosamples per second. However, in the case of G.728, five of these samples are grouped and transmitted as a single vector using conventional vector quantization. The vector codebook has 1,024 entries, requiring a 10-bit code to be transmitted at the rate of 1,600 words per second for a total transmitted bit rate of 16 Kbps. Each of the 1,024 possible excitation signals is evaluated to find the one that minimized the mean square error in a perceptually weighted space. The coder uses LPC analysis to create three different linear predictors: a fiftieth-order prediction filter for the next sample value, a tenth-order filter that guides the quantizer step size, and a perceptual weighting filter that is used to select the excitation signal.

This coder is considered to be toll quality. In combination with G.711 input and output, this coder is considered to be equivalent to 3.5 QDU. The coder was extensively tested, including low-level and high-level inputs and was shown to behave similarly to other network quality standards such as G.721 32-Kbps ADPCM. The coder is remarkably robust to random bit errors. At 10^{-3}

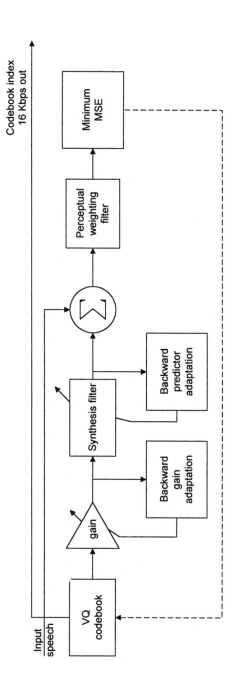

Figure 4.6 G.728 encoder.

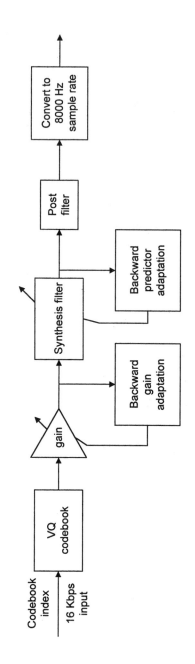

Figure 4.7   G.728 decoder.

random bit errors, the coder is significantly more robust than any other ITU-T coder—the effects are barely perceptible. It also performs better at 10^{-2} BER than any other speech coder, but the effects are more noticeable. There was no ITU-T requirement for frame erasure concealment in the original terms of reference. Because of the interest in using this coder for mobile applications, work is now under way to modify the decoder to provide frame erasure concealment.

The coder has been tested in more than a half-dozen languages and with various types of background noises and with music. No language dependency was found and no annoying artifacts were produced with nonspeech inputs. The coder has only been formally tested by the ITU-T over IRS microphone characteristics and telephone handsets for listening. It has been tested at AT&T with non-IRS flat-weighted speech and with headphone listening. It has always scored as well as the 32-Kbps ADPCM standard in these tests. It has also been played over loudspeakers many times. Many listeners have commented that it sounds better over a loudspeaker than 32-Kbps ADPCM.

Two tandem encodings are considered equivalent to 7 QDU. The coder was originally weakest in this area and revisions were made, which delayed the standard by one year. The result was the current performance, which actually exceeded the goals set in the terms of reference by matching the performance of 32-Kbps ADPCM for up to four tandem encodings.

The coder has approximately a 2-ms one-way delay if implemented as described in the specification. It is possible to implement the coder in 2.5-ms blocks, which would result in a one-way delay of 7.5 to 10 ms, depending on other aspects of the processor and the system.

The floating-point recommendation for G.728 includes an appendix with implementation test vectors. Many laboratories have created C or Fortran language simulations that have passed the test vectors. DSP implementations based on standard floating-point DSP chips exist that have successfully passed the test vectors. A bit exact fixed-point specification was created that is fully interoperational with any floating-point implementation that passes the floating-point test vectors.

G.728 is a high complexity algorithm. In floating point it takes 16 to 18 MIPS. A fixed-point specification was finalized in 1994. The combined encoder and decoder have been implemented in as few as 27 MIPS in fixed point using just under 2,048 words of RAM.

4.2.5   G.729 (8 Kbps, CS-ACELP)

In response to a request from CCIR Task Group 8/1 for a speech coder for wireless networks as envisioned in the mobile networks, the ITU initiated a work program for a toll quality 8-Kbps speech coder in 1990. The original delay requirements were relaxed in 1991 to reduce the expected complexity of the coder. In 1993 and early 1994, two potential coders were tested. Both coders met the performance requirements for clean input speech quality. However, both failed the requirements for other input levels, for frame erasure concealment, and for speech with noisy backgrounds or music. In 1994, an agreement was reached that these coders would be combined and that their proponents would collaborate to produce an optimized coder capable of meeting all requirements.

In 1995, testing was repeated by four subjective test labs in four different languages. SQEG concluded that on the basis of these test results, the optimized coder met all speech quality requirements. Further characterization testing in late 1995 agreed with this conclusion. However, not all the tests were in agreement about the quality of speech with acoustic background noise. The more sensitive tests, such as the comparison category rating (CCR) test, indicated that there was lesser quality for G.729 than for the G.726 reference codec. However, it was the opinion of the SQEG that the test methodology was not well enough understood to accept these results. Absolute category rating (ACR) test results contradicted the CCR results. The scores for G.729 were not significantly different than those of G.726 and were sometimes higher for noisy background conditions.

G.729 is a linear prediction analysis-by-synthesis coder. The whole name is conjugate structure algebraic codebook excitation linear prediction. The conjugate structure is in the quantization of the linear prediction coefficients. The algebraic codebook refers to the partitioning of the excitation space. This allows for a fast, almost optimal search of all possible excitation sequences. Two subframes of 5 ms each are combined with a single estimate of the pitch period and gain to form the excitation. The LPC filter is updated every 10 ms and is quantized in the line spectral pair domain using the conjugate quantization. G.729 has a 10-ms frame size and 5-ms look-ahead. This results in a 35-ms one-way delay when computation, algorithmic delay, and transmission are included. The complexity is reported to be about 15 MIPS and about 3,300 words of RAM are required. This coder and G.723.1 are the first ITU coders to be specified by a bit-exact fixed-point ANSI C code simulation of the encoder and decoder. This recommendation was put forward for determination by SG15 in November 1995 and became an official ITU recommendation in 1996.

The basic version of G.729 is relatively complex. The ITU approved annex A to G.729, which is less complex, in May 1996. As one would expect, the performance of annex A is inferior to the basic system.

4.2.6   G.723.1 (5.3 Kbps/6.3 Kbps, ACELP/MP-MLQ)

In September 1993, the ITU established a project to develop a multimedia videophone standard for operation over the public-switched telephone network and mobile networks. This program, led by the author, was highly successful, resulting in the H.324 suite of recommendations. To solve this very challenging problem, it was necessary to develop a speech coding standard to provide toll quality at the lowest possible bit rate using technology which was available at that time period. The speech codec project, led by Richard Cox, was also highly successful, resulting in Recommendation G.723.1.

The G.723.1 standard specifies two different transmission bit rates, 5.3 and 6.3 Kbps, both of which are mandatory. When operating at the higher rate, the coder provides toll quality, while the quality at the low bit rate is slightly reduced, ideal for mobile applications. G.723.1, like G.729, employs a basic ACELP algorithm. The primary difference between the two G.723.1 coders is the excitation codebook. The 6.3-Kbps coder has a multipulse excitation, giving rise to the designation *MultiPulse-Maximum Likelihood Quantizer* (MP-MLQ), while the 5.3-Kbps coder employs the conventional ACELP excitation.

The G.723.1 standard, operating at 6.3 Kbps, achieves the same quality as G.729 operating at 8 Kbps for two reasons: (1) the multipulse excitation codebook, (2) the greater delay—30 ms rather than 10 ms. In general, if a coder has more time to compress a signal, the compression should be greater. The G.723.1 standard is used exclusively in the H.324 standard and is the de facto standard for voice over Internet where the bit rate is critical. More information on G.723.1 is provided in Section 8.3.

4.2.7 G.WSC (Wideband Speech Coder)

For a number of years, the ITU has been working on a new wideband speech (7-kHz bandwidth) coder to update the old technology embodied in G.722. The objective is to develop a coder primarily for the videoconferencing application—therefore having low bit rate, low delay, error resilience, and very good speech quality. The target bit rates are 16, 24, and 32 Kbps.

4.2.8 4.0-Kbps Speech Coder

The ITU has established a project to develop a new speech coder to provide toll speech quality at 4 Kbps. The terms of reference of this system are based, to a large extent, on the terms for reference used by the ITU 16- and 8-Kbps speech coding efforts, with minor modifications where appropriate.

The applications identified for the 4-Kbps speech coding algorithm are partitioned into two groups: a primary group and a secondary group. The primary group comprises those applications that are most likely to employ 4-Kbps speech coding early and in large numbers. As a result, primary applications are expected to "drive" the development of the standard, at least as regards schedule. The secondary group comprises those applications likely to benefit from the availability of a 4-Kbps unique speech coding standard, but which are either unlikely to employ large numbers of 4-Kbps speech coding devices or, at least on an interim basis, can also utilize some other speech coding standards without adversely impacting the economics of their application as shown in Table 4.2.

A few highlights of the schedule for this program are listed in Table 4.3.

4.3 MPEG Audio

While the primary objective of the ITU G.7XX standards is to provide *speech* quality equivalent to the telephone network, the key objective of the MPEG audio standards is to provide high *audio* quality for broadcast applications, including, of course, music as one of the most challenging audio signals. *Compact disc* (CD) audio is a point of reference for MPEG audio coding. A CD recording employs a bandwidth from 10 Hz to 22 kHz, samples the audio at 44.1 kHz, encodes each sample with 16 bps for a bit rate of 705.6 Kbps. For stereo recording this yields an overall bit rate of 1.41 Mbps.

The goal of the MPEG 1 audio standard is to achieve transparent, or CD quality, for the stereo application. The MPEG 1 standard has three parts, or layers, capable of delivering increasing quality as the number of layers is increased. If only the first layer is employed, transparent quality is achieved at 384 Kbps. If two layers are used, the bit rate can be reduced to 256 Kbps.

The objective of MPEG 2 audio is to create a multichannel movie theater-style sound system consisting of 5 channels of audio as illustrated in Figure 4.8. There are two MPEG 2 audio standards, one fully compatible with MPEG 1, and one which is not backward compatible (NBC). The compatible standard provides transparent quality at a channel bit rate of 128 Kbps, while the NBC standard reduces this channel bit rate to 64 Kbps.

Table 4.2
Applications for 4-Kbps Speech Coder

Primary Applications	Secondary Applications
Very low-rate PSTN visual telephony	Digital circuit multiplication
Personal communications	Packet circuit multiplication equipment
Simultaneous voice and data systems	Low-rate mobile visual telephony
Mobile-telephony satellite systems	Message retrieval systems
	Private networks

Table 4.3
Schedule for 4-Kbps Speech Coder

Sept. 1998	- Define test plan for the selection phase - Possible reduction of the number of candidates on the basis of the capability to meet performance requirements
Jan. 1999	- Fixed-point C-simulation available - Commence selection testing
June 1999	- Selection of one candidate for further testing - Commence drafting of final text and C-code
Oct. 1999	- Commence characterization testing
Nov. 1999	- "Determination" of Recommendation and C-source code

The basic technology used in MPEG audio codes, known as auditory masking, is presented herein. In a quiet environment, there is a nonuniform sound pressure threshold as a function of frequency above which the sound will be audible to a human. This threshold is called the *threshold in quiet*. When a signal such as a tone is applied, the sound pressure threshold that must be exceeded by other sounds to be audible is changed. This new threshold is called the *masking threshold*. In other words, in frequency domain masking, a faint signal may be completely masked if it is in the vicinity of louder signals with similar frequency content. The masking process draws upon human auditory perception, which performs what is called a *critical band analysis* in the inner ear. This critical band

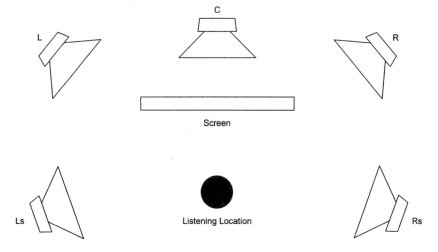

Figure 4.8 Speaker configuration for multichannel MPEG 2 audio.

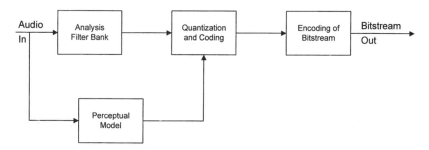

Figure 4.9 Basic structure of a perceptual coding scheme.

analysis in the auditory system can be modeled as a bandpass filter bank with bandwidths becoming progressively larger with increasing frequency. Temporal masking is the idea that a particular signal can mask other signals that occur just before (premasking) and just after (postmasking) the stronger signal. Premasking has a much shorter possible interval than postmasking, but both are significant and are exploited. The basic approach to perceptual coding is illustrated by the block diagram in Figure 4.9. The input signal is processed first by an analysis filterbank, also known as a subband filter, which translates the input into the frequency domain. Using rules from psychoacoustics, a perceptual model estimates the masking thresholds. This information is passed to the quantizer so that all quantization noise is below the masking thresholds. The output coded bitstream consists of the encoded quantized data plus some side information, such as the bit allocation scheme.

5

Audiovisual Standards Organizations

It has been clearly established that standards are a fundamental necessary step in the provision of effective audiovisual communications. The Group 3 facsimile and H.320 videoconferencing standards are excellent examples. Figure 5.1 illustrates the primary international and U.S. organizations that establish these standards. Section 5.1 focuses on the international standard organizations such as the ITU, ISO, IMTC, SMPTE, IEEE, and IETF. It should be noted that the ITU and ISO organizations are fundamental standard-setting structures in that the basic membership is comprised of countries. In contrast, the IMTC is a consortium of corporations. T1 and NCITS are two key U.S. standard organizations dealing with digital TV standards which are discussed in Section 5.2. The European Telecommunications Standards Institute (ETSI) is a regional standards-setting organization described in Section 5.3.

5.1   International Standards Organizations

5.1.1   International Telecommunications Union

5.1.1.1   Mission, Structure, and Operation of the ITU

The ITU is a global intergovernmental treaty organization within which governments and the private sector cooperate to develop recommendations that guide telecommunication service providers and manufacturers of related equipment to offer interoperable telecommunication services. The ITU also allocates radio-frequency spectrum and supports efforts to eliminate harmful interference between radio stations of different specific technical assistance projects funded by the United Nations. These general task outlines are pursued by the ITU via

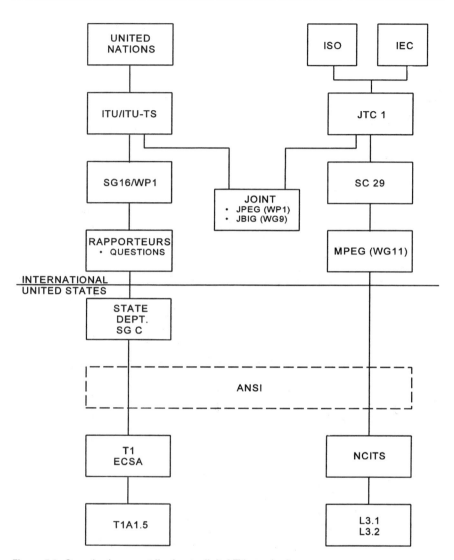

Figure 5.1 Organizations contributing to digital TV standards.

a three-sector organizational structure reflected in Figure 5.2. Each sector is supported by an ITU bureau headed by a director with general administration managed by a general secretariat all under the secretary-general of the ITU.

The method of operation in the ITU-TS is by way of consensus. The concept of consensus should be understood in terms of its practice in the ITU. The view most often supported is that consensus is arrived at when no member objects to the proposal at hand or when only one or two members wish to indicate their objections to some aspect of the proposal but do not resist it as a

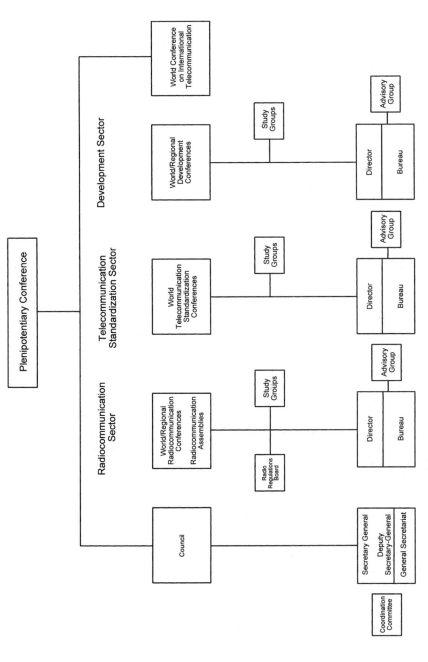

Figure 5.2 Organizational structure of the ITU.

whole. The most difficult scenario is the one in which one or two members object to the proposal in its entirety. In such a case, chairman decision and voting are not appropriate mechanisms at Working Party or Study Group meetings except when the recommendation approval process is in progress at a Study Group meeting. No one suggests that ITU consensus is the same as T1 consensus or consensus as it might be defined elsewhere. The success of the ITU to date may be at least partially attributed to the use of consensus as practiced therein.

A few years ago, the ITU was organized on a rigid four-year work period such that standards were established every four years. They were published in sets of, for example, "Red" books and "Blue" books. It was determined that this process was too slow, and the Resolution 1 process was instituted that permits standards to be established at any time. The Resolution 1 process specifies that a recommendation is established using a three-step sequence. First, a Study Group "determines" that a draft recommendation is sufficiently stable to give notice of the intention to invoke the Resolution 1 process. At a subsequent meeting, the Study Group takes the second step of "deciding" to invoke the Resolution 1 process. Finally, the draft recommendation is submitted to membership countries for ballot.

5.1.1.2 Telecommunication Standardization Sector

The Telecommunication Standardization Sector (ITU-T) was created on March 1, 1993, within the framework of the "new" ITU, along with the Radiocommunications Sector (ITU-R) and the Telecommunication Development Sector (ITU-D). The ITU-T replaces the former CCITT.

The purpose of the ITU-T is the elaboration, adoption, distribution, and follow-up of recommendations (nonbinding standards) to standardize telecommunications on a worldwide basis. These recommendations are developed from the study of questions (that is, areas for study) for international telecommunications, in 15 Study Groups (SGs), by some 4,000 experts from throughout the world.

The standardization process is cyclic; experts elaborate draft standard texts, discuss them during periodic Study Group meetings, amend them after discussion, present revised text to the next meeting, and so on, until the Study Group considers the text to be mature enough to be presented to the Members (Administrations) or ITU for approval as an ITU-T recommendation. This approval is done in most cases by correspondence.

By right, all ITU Members may participate in ITU-T. Other entities and organizations, with the approval of the Administrations, can become "members" (with small *m*'s), and participate as well.

Such entities and organizations include:

- Network or service providers, manufacturers or suppliers, financial or development institutions;
- Regional organizations and other international telecommunication, standardization, financial, or development organizations;
- Other entities dealing with telecommunication matters.

As indicated earlier, the high-level ITU organization responsible for setting standards, in the videoconferencing and videophone areas, is the ITU-T. The ITU-T, in turn, is divided into fifteen Study Groups, which are listed in Table 5.1. The work performed by the ITU is defined in terms of questions that define the standards to be developed. The ITU assigns a "rapporteur" to be responsible for the work performed for each of the questions, and each rapporteur assembles a group of experts to perform the work. The work is performed by experts submitting technical contributions for review by the Experts Group and by the supervisory Study Group.

The ITU-T structure comprises:

- The World Telecommunication Standardization Conference (WTSC), which defines general policy for the sector, approves the program of work through the questions for study, and allocates them to the study groups;
- The Telecommunication Standardization Advisory Group (TSAG), which looks at work priorities and strategies for the sector, follows up on accomplishments of the work program, and advises the TSB Director and the study groups;
- The telecommunication standardization study groups and their working parties (WPs), which bring together the experts who study the questions and elaborate the recommendations;
- The Telecommunication Standardization Bureau (TSB) which provides the sector secretariat and coordinates the standardization process.

5.1.1.3 World Telecommunication Standardization Conference

The WTSC normally meets every four years to examine matters that are specific to telecommunication standardization, adopted by established procedures, or posed by the Plenipotentiary Conference (supreme organ of ITU) or by the Council (executive organ of the Plenipotentiary Conference).

The WTSC is responsible for the work program, working methods, and structure of the Study Groups and the procedures for approving recommendations. It examines Study Group reports:

Table 5.1
Telecommunication Standardization Sector (ITU-T)

Study Group	Title	Recommendations
2	Network and service operation	
3	Tariff & accounting principles including related telecommunications economic & policy issues	
4	TMN & network maintenance	
5	Protection against electromagnetic environment effects	
6	Outside plant	
7	Data networks & open system communications	
8	Characteristics of telematic systems	Facsimile—T.3, T.30
9	Television & sound transmission	
10	Languages & general software aspects for telecommunication systems	
11	Signaling requirements & protocols	
12	End-to-end transmission performance of networks & terminals	
13	General network aspects	I-SERIES (ISDN)
15	Transport networks, systems & equipment	
16	Multimedia services & systems	
1	Modems	V.8, V.32, V.34, V.PCM
2	Multimedia systems	H.3XX, H.2XX, T.12X, T.13X
3	Signal coding	Speech Coding (G.7XX)
		Video Coding (H.26X)

- Establishes the work program resulting from examining the questions, according to their urgency and priority, and evaluates financial impact and establishes a schedule to complete the corresponding studies;

- Decides whether to maintain or dissolve existing study groups or to create new ones and fixes their work program;

- Supervises the activities of the TSB and examines and approves the Director's report on sector activities since the preceding conference;

- Approves, modifies, or rejects recommendations contained in the reports, although most of them are approved by correspondence.

The first WTSC called within the new ITU framework took place in Helsinki in 1993. It provided the questions for study by the appropriate Study Groups during the new study period (the period between two successive conferences), that is, for the years 1993 to 1996.

5.1.1.4 Telecommunication Standardization Advisory Group

Due to the constant and rapid evolution of telecommunications, the WTSC constituted the TSAG to:

- Study the priorities and strategies to adopt in the framework of ITU-T activities;
- Examine the progress in the execution of the work program for the sector;
- Formulate recommendations with regard to the work of the Study Groups;
- Recommend measures to help the cooperation and coordination between ITU-T, the other ITU Sectors, and the ITU Strategic Planning Unit.

The TSAG follows the work of ITU-T, identifies changing requirements, and provides advice on appropriate changes, giving due regard to the cost and availability of resources. The TSAG also analyzes sector working methods and proposes possible ways of improving these methods.

To fulfill its mission, the TSAG has working groups to deal with:

- Computerized work program and priorities for Study Groups;
- Work programs external to ITU and projected results, so as to harmonize the ITU-T activities with those of other standardization organization;
- Follow-up on work progress within Study Groups;
- Promotion of electronic document handling.

5.1.1.5 Telecommunication Standardization Bureau

A Director, elected by the ITU Plenipotentiary Conference, directs the TSB and organizes and coordinates the activities of the ITU-T.

The Bureau:

- Provides the necessary administrative, logistic, and technical support to World Conferences and to telecommunication standardization Study Groups and their working parties;

- Applies the provisions of the International Telecommunication Regulations;

- Coordinates the elaboration, editing, and distribution of circulars, documents, and publications for the sector;

- Facilitates the exchange of administrative, operational, statistical, and tariff information among the agencies operating public international telecommunication services, notably through the ITU Operational Bulletin;

- Provides technical information on international telecommunications and collaborates closely with both the ITU-D, for matters of interest to developing countries, and the ITU-R.

The ITU's Telecommunication Standardization Bureau can be reached at the following addresses and numbers:

Street address:	CH 1211 Geneva 20, Switzerland
Telephone:	+41 22 730 5852
Facsimile:	+41 22 730 5853
Telex:	421 000 uit ch
Internet email:	tsbmail@itu.ch
X.400email:	s=tsbmail; P=itu; A=arcom; C=ch

5.1.2 International Standards Organization

The ISO is a worldwide federation of national standards bodies from some 100 countries, one from each country. ISO is a nongovernmental organization established in 1947. The mission of the ISO is to promote the development of standardization and related activities in the world with a view to facilitating the international exchange of goods and services and to developing cooperation in the spheres of intellectual, scientific, technological, and economic activity. ISO's work results in international agreements, which are published as international standards.

5.1.2.1 How It All Started

International standardization began in the electrotechnical field; the International Electrotechnical Commission (IEC) was created in 1906. Following a meeting in London in 1946, delegates from 25 countries decided to create a new international organization "the object of which would be to facilitate the international coordination and unification of industrial standards." The new organization, ISO, began to function officially on February 23, 1947.

5.1.2.2 International Standardization: What Does It Achieve?

Industry-wide standardization is a condition existing within a particular industrial sector when the large majority of products or services conform to the same standards. It results from consensus agreements reached between all economic players in that industrial sector, such as suppliers, users, and often governments. They agree on specifications and criteria to be applied consistently in the choice and classification of materials, the manufacture of products, and the provision of services. The aim is to facilitate trade, exchange, and technology transfer through:

- Enhanced product quality and reliability at a reasonable price;
- Improved health, safety and environmental protection, and reduction of waste;
- Greater compatibility and interoperability of goods and services;
- Simplification for improved usability;
- Reduction in the number of models and thus reduction in costs;
- Increased distribution efficiency and ease of maintenance.

Users have more confidence in products and services that conform to international standards. Assurance of conformity can be provided by manufacturers' declarations or by audits carried out by independent bodies.

5.1.2.3 Why Is International Standardization Needed?

The existence of nonharmonized standards for similar technologies in different countries or regions can contribute to so-called technical barriers to trade. Export-minded industries have long sensed the need to agree on world standards to help rationalize the international trading process. This was the origin of the establishment of ISO.

International standardization is now well-established for very many technologies in such diverse fields as information processing and communications, textiles, packaging, distribution of goods, energy production and utilization, shipbuilding, banking, and financial services. It will continue to grow in importance for all sectors of industrial activity for the foreseeable future.

5.1.2.4 Who Makes Up ISO?

A member body of ISO is the national body "most representative of standardization in its country." It follows that only one such body for each country is accepted for membership. The member bodies have four principal tasks:

1. Inform potentially interested parties in their country of relevant international standardization opportunities and initiatives;

2. Organize so that a concerted view of the country's interests is presented during international negotiations leading to standards agreements;

3. Ensure that a secretariat is provided for those ISO technical committees and subcommittees in which the country has an interest;

4. Provide their country's share of financial support for the central operations of ISO, through payment of membership dues.

5.1.2.5 Who Does the Work?

The technical work of ISO is highly decentralized, carried out in a hierarchy of some 2,700 technical committees, subcommittees, and working groups. In these committees, qualified representatives of industry, research institutes, government authorities, consumer bodies, and international organizations from all over the world come together as equal partners in the resolution of global standardization problems.

The major responsibility for administrating a standards committee is accepted by one of the national standards bodies that make up the ISO membership (for example, AFNOR, ANSI, BSI, CSBTS, DIN, and SIS). The member body holding the secretariat of standards committee normally appoints one or two persons to conduct the technical and administrative work. A committee chairman assists committee members in reaching consensus. Generally, consensus will mean that a particular solution to the problem at hand is the best possible one for international application at that time.

The Central Secretariat in Geneva acts to ensure the flow of documentation in all directions, to clarify technical points with secretariats and chairmen, and to ensure that the agreements approved by the technical committees are edited, printed, submitted as draft international standards to ISO member bodies for voting, and published. Meetings of technical committees and subcommittees are convened by the Central Secretariat, who coordinates all such meetings with the committee secretariats before setting the date and place. Although the greater part of the ISO technical work is performed via correspondence, there are, on average, a dozen ISO meetings taking place somewhere in the world every working day of the year. Each member body interested in a subject has the right to be represented on a committee. International organizations, governmental and nongovernmental, in liaison with ISO, also take part in the work. ISO collaborates closely with the IEC on all matters of electrotechnical standardization.

The scope of ISO is not limited to any particular branch; it covers all standardization fields except electrical and electronic engineering, which is the responsibility of IEC. The work in the field of information technology is carried out by a joint ISO\/IEC technical committee (particularly JTC 1).

5.1.2.6   How Are ISO Standards Developed?

ISO standards are developed according to the following principles:

- *Consensus:* The views of all interests—such as manufacturers, vendors and users, consumer groups, testing laboratories, governments, engineering professions, and research organizations—are taken into account.
- *Industrywide:* Global solutions are sought to satisfy industries and customers worldwide.
- *Voluntary:* International standardization is market-driven and therefore based on voluntary involvement of all interests in the market price.

There are three main phases in the ISO standards development process.

The need for a standard is usually expressed by an industry sector, which communicates this need to a national member body. The latter proposes the new work item to ISO as a whole. Once the need for an international standard has been recognized and formally agreed upon, the first phase involves defining the technical scope of the future standard. This phase is usually carried out in working groups that are comprised of technical experts from countries interested in the subject matter.

Once agreement has been reached on which technical aspects are to be covered in the standard, a second phase is entered during which countries negotiate the detailed specifications within the standard. This is the consensus-building phase.

The final phase comprises the formal approval of the resulting draft international standard (the acceptance criteria stipulate approval by two-thirds of the ISO members that have participated actively in the standards development process and approval by 75% of all members that vote), following which the agreed text is published as an ISO international standard.

5.1.2.7   Searching for Information

Enquiries about standards involve those of ISO and a number of recognized standards agreed within other international technical organizations. There are, in addition, several hundred thousand standards and technical regulations in use throughout the world containing special requirements for a particular country or region. Finding information about all these standards, technical regulations, or related testing and certification activities can be a heavy task.

The ISO Information Network (ISONET) is there to ease the problem. This is a worldwide network of national standards information centers that have cooperatively developed a system to provide rapid access to information about standards, technical regulations, and testing and certification activities currently used in different parts of the world. Members of this network—usually the ISO member for any given country—act effectively as "specialized enquiry points" in the dissemination of information and in identifying the relevant sources of information for solving specific problems. The Agreement on Technical Barriers to Trade, drawn up under the General Agreement on Tariffs and Trade (GATT) (that is, the future WTO), calls upon its signatories to establish in each country an enquiry point capable of answering questions about the standards, technical regulations, and certification systems in force in that country. In many countries, the ISONET enquiry point and the GATT enquiry point are one and the same.

5.1.3   International Multimedia Teleconferencing Consortium

The International Multimedia Teleconferencing Consortium (IMTC) is a non-profit corporation composed of, and supported by, more than 140 organizations from North America, Europe, Asia, and the Pacific. The IMTC's mission is to promote, encourage, and facilitate the development and implementation of interoperable multimedia teleconferencing solutions based on open international standards. In this regard, the IMTC is currently focused on multimedia teleconferencing standards adopted by the ITU, specifically the ITU-T, T.120, H.320, H.323, and H.324 standards.

Key activities include sponsoring and conducting interoperability test sessions between suppliers of conferencing products and services based on these standards; educating the business and consumer communities on the status, value, and benefits of the underlying technologies and resultant applications; and encouraging IMTC members to make submissions to standards bodies which enhance the interoperability and usability of multimedia teleconferencing products and services.

The IMTC was formed in September 1994 through a merger of two other organizations, the Consortium for Audiographics Teleconferencing Standards (CATS) and the Multimedia Communications Community of Interest (MCCOI). CATS, formed in 1993, had focused on the T.120 standards suite for audiographics teleconferencing. MCCOI, also formed in 1993, had focused more on the ITU H.320 standards suite for video telephony.

In December 1995, IMTC merged with the Personal Conferencing Work Group (PCWG). PCWG, which represented 120 Members and Observers worldwide, had also focused on multimedia teleconferencing standards as well as the needs of users of products and services in this category.

The IMTC's 140-plus member organizations work together to facilitate the implementation of teleconferencing products and services based on the T.120, H.320, H.323 and H.324 standards; recommend enhancements and develop specifications that improve these standards; assure interoperability through their participation in industry-wide IMTC sponsored test sessions; and educate the marketplace on trends, developments, and progress in the field.

The IMTC is structured as an open organization, and is controlled by its Voting Members. IMTC's Board of Directors, which oversees the management of the corporation and establishes its operating policies, is elected annually by the Voting Members. Any Voting Member can submit a candidate in these elections. Voting Members also ratify IMTC's annual budget and any changes to IMTC's corporate by-laws. Their rights and privileges are specified in IMTC's by-laws, which ensures that any changes to rights and privileges are ratified by a majority.

The IMTC also follows a defined consensus decision making process. Specifications and recommendations which would require product development work, and issues which commit the IMTC to a specific course of action, are balloted within the membership for majority approval. All Members, regardless of their size, are encouraged to participate, voice opinions, and help resolve issues. This reflects IMTC's strong commitment to an inclusive, open structure and a "make things happen" environment.

The IMTC is focused on interoperability. Four of IMTC's eight activity groups are involved in the planning and implementation of interoperability test sessions for products and services based on the T.120, H.320, H.323 and H.324 standards. These efforts will help minimize market fragmentation, and encourage prospective customers to implement and use multimedia teleconferencing solutions on the Internet, corporate intranets, and the PSTN more rapidly.

5.1.3.1 Activities and Initiatives

To achieve its objectives, the IMTC maintains a Technical Steering Committee and eight activity groups. These activity groups and their activities are summarized below.

- *Data Conferencing:* This group is chartered with the development of plans, methodologies, and facilities for T.12x and T.130 standards interoperability testing.

- *Interworking and Networking Services:* This activity group works to address issues concerning service provisioning, network management, and cross-standard implementation and interoperability.

- *Packet Network Conferencing:* This activity group is responsible for extending existing H.323 standards as needed to expedite LAN/WAN telephony, videoconferencing, and multimedia collaboration.

- *Switched Network Conferencing:* This activity group works to develop plans, methodologies, and facilities for H.320 and H.324 interoperability testing.

- *Voice Over IP (VoIP) Forum:* The VoIP Forum is responsible for defining and establishing guidelines for voice and voiceband data traffic over IP data networks. The group has six working committees, each focusing on a particular industry protocol to establish and recommend technical guidelines for VoIP technologies.

- *Network Transport End-to-End Quality of Service:* This recently added activity group examines existing efforts from other standard bodies such as the IEFT and ATM Forum. The groups makes recommendations for change to drive solutions across disparate networks.

- *User, Usability, and Applications:* The new activity group is responsible for examining and addressing usability and applications issues at all levels of an end-to-end network. The group oversees the User Advisory Council, a group of conferencing that provides feedback to the IMTC.

- *Marketing:* This group is responsible for marketing the IMTC mission to the media, industry analysts, and other communities of interest; educating end-users on the value of and progress towards interoperable products and services; and incorporating and communicating market needs to the other activity groups. Current initiatives include an ongoing public relations program, development of relevant collateral and position papers, implementing IMTC presence at industry events, supporting the interoperability programs, and collecting, maintaining, and distributing industry information.

In addition to the above, the IMTC maintains formal liaison with the ITU Study Group 8, ITU Study Group 15, ISO/IEC JTC 1, DAVIC, and EV. These relationships ensure that the IMTC is able, through its members and as an organization, to monitor developments of relevance to its mission and provide the appropriate direction, input, and guidance.

5.1.4 Society of Motion Picture and Television Engineers

The Society of Motion Picture and Television Engineers (SMPTE) is an international technical society devoted to advancing the theory and application of motion-imaging technology including film, television, video, computer imaging, and telecommunications. The SMPTE currently serves 8,500 members in 72 countries. Members of the Society are engineers, executives, technical directors, cameramen, editors, consultants, and specialists in film processing and film and television production and post-production; and practitioners from almost every other discipline in the motion-imaging industry.

The Society was founded in 1916, as the Society of Motion Picture Engineers. The *T* was added in 1950 to embrace the emerging television industry. The SMPTE is recognized around the globe as a leader in the development of standards and authoritative, consensus-based recommended practices (RPs) and engineering guidelines (EGs). The Society serves all branches of motion imaging including film, video, and multimedia.

To accomplish its educational goals, the SMPTE organizes annual conferences as well as select educational forums, including seminars and tutorials. It also publishes the *SMPTE Journal*, recognized around the world for its technical papers, tutorials, articles on practical applications, standards updates, and reports on SMPTE Sections' activities.

Industry professionals volunteer on various SMPTE technology committees to develop standards, RPs, and EGs. The standardization efforts of the Society have been recognized with several national and international awards, including three Emmy awards and an Oscar. The Oscar was won in 1957 for the Society's advancement of the motion picture industry. The three Emmy awards include one in 1983 for the development of the 4:2:2 digital component video standard, another in 1986 for the development of the D-1 recording system standards, and still another in 1993 for in plant digital serial interconnection for television standards.

5.1.5   Institute of Electronic and Electrical Engineers

As the world's largest professional technical society, the Institute of Electronic and Electrical Engineers (IEEE) has more than 315,000 members in over 150 countries.

IEEE Standards is a world leader in the development and dissemination of voluntary, consensus-based industry standards involving today's leading edge electrotechnology. The organization is committed to broadening the scope of the standards products and services it produces to embrace the demands of new technologies and to respond to global market needs. IEEE Standards supports international standardization and encourages the development of globally acceptable standards.

IEEE promotes the development of electrotechnology and allied sciences and the application of those technologies for the benefit of humanity, the advancement of the profession, and the well-being of its members.

5.1.6 Internet Engineering Task Force

The Internet Engineering Task Force (IETF) is a loosely self-organized group of people who make technical and other contributions to the engineering and evolution of the Internet and its technologies. It is the principal body engaged in the development of new Internet standard specifications. Its mission includes:

- Identifying, and proposing solutions to, pressing operational and technical problems in the Internet;
- Specifying the development or usage of protocols and the near-term architecture to solve such technical problems for the Internet;
- Making recommendations to the Internet Engineering Steering Group (IESG) regarding the standardization of protocols and protocol usage in the Internet;
- Facilitating technology transfer from the Internet Research Task Force (IRTF) to the wider Internet community;
- Providing a forum for the exchange of information within the Internet community between vendors, users, researchers, agency contractors, and network managers.

The IETF meeting is not a conference, although there are technical presentations. The IETF is not a traditional standards organization, although many specifications are produced that become standards. The IETF is made up of volunteers who meet three times a year to fulfill the IETF mission.

There is no membership in the IETF. Anyone may register for and attend any meeting. The closest thing there is to being an IETF member is being on the IETF or working group mailing lists (see the IETF mailing lists section). This is where the best information about current IETF activities and focus can be found.

5.1.6.1 Organization

To completely understand the structure of the IETF, it is useful to understand the overall structure in which the IETF resides. There are four groups in the structure: the Internet Society (ISOC) and its Board of Trustees, the Internet Architecture Board (IAB), the IESG, and the IETF itself.

The ISOC is a professional society that is concerned with the growth and evolution of the worldwide Internet, with the way in which the Internet is and can be used, and with the social, political, and technical issues which arise as a result. The ISOC Trustees are responsible for approving appointments to the IAB from among the nominees submitted by the IETF nominating committee.

The IAB is a technical advisory group of the ISOC. It is chartered to provide oversight of the architecture of the Internet and its protocols, and to serve, in the context of the Internet standards process, as a body to which the decisions of the IESG may be appealed. The IAB is responsible for approving appointments to the IESG from among the nominees submitted by the IETF nominations committee.

The IESG is responsible for technical management of IETF activities and the Internet standards process. As part of the ISOC, it administers the process according to the rules and procedures which have been ratified by the ISOC Trustees. The IESG is directly responsible for the actions associated with entry into and movement along the Internet "standards track," including final approval of specifications as Internet standards.

The IETF is divided into eight functional areas: applications, Internet, IP: next generation, network management, operational requirements, routing, security, and transport and user services. Each area has one or two area directors. The area directors, along with the IETF/IESG Chair, form the IESG.

Each area has several working groups. A working group is a group of people who work under a charter to achieve a certain goal. That goal may be the creation of an informational document, the creation of a protocol specification, or the resolution of problems in the Internet. Most working groups have a finite lifetime. That is, once a working group has achieved its goal, it disbands. As in the IETF, there is no official membership for a working group. Unofficially, a working group member is somebody who is on that working group's mailing list; however, anyone may attend a working group meeting.

Areas may also have Birds of a Feather (BOF) sessions. They generally have the same goals as working groups, except that they have no charter and usually only meet once or twice. BOFs are often held to determine if there is enough interest to form a working group.

5.1.6.2 IETF Mailing Lists

Anyone who plans to attend an IETF meeting should join the IETF announcement mailing list. This is where all of the meeting information, Internet draft and request for comment (RFC) announcements, and IESG Protocol Actions and Last Calls are posted. People who would like to "get technical" may also join the IETF discussion list. The email address is: ietf@ietf.org. This is where discussions of cosmic significance are held (most working groups have their own mailing lists for discussions related to their work). To join the IETF announcement list, send a request to: ietf-announce-request@ietf.org
To join the IETF discussion list, send a request to: ietf-request@ietf.org

5.1.6.3 RFCs and Internet Drafts

Originally, RFCs were just what the name implies: requests for comments. The early RFCs were messages between the ARPANET architects about how to resolve certain problems. Over the years, RFCs became more formal. It reached the point that they were being cited as standards, even when they weren't.

To help clear up some confusion, there are now two special subseries within the RFCs: For Your Information (FYIs) and STDs. The RFC subseries was created to document overviews and topics that are introductory. Frequently, FYIs are created by groups within the IETF User Services Area. The STD RFC subseries was created to identify those RFCs that do in fact specify Internet standards.

Every RFC, including FYIs and STDs, has an RFC number by which it is indexed and by which it can be retrieved. FYIs and STDs have FYI numbers and STD numbers, respectively, in addition to RFC numbers. This makes it easier for a new Internet user, for example, to find all the helpful, informational documents by looking for the FYIs among all the RFCs. If an FYI or STD is revised, its RFC number will change, but its FYI or STD number will remain constant for ease of reference.

5.2 U.S. Standards Organizations

5.2.1 Standards Committee T1—Telecommunications

5.2.1.1 Introduction

Incorporated as a not-for-profit association in 1983, the Exchange Carriers Standards Association (ECSA) became the Alliance for Telecommunications Industry Solutions (ATIS) in 1993. ATIS consists of and is supported by members of the telecommunications industry to address exchange access, interconnection, and other technical issues in a postdivestiture telecommunications era. ATIS sponsors and provides secretariat support to Committee T1—Telecommunications Committee accredited by the ANSI. ATIS sponsors a variety of postdivestiture industry forums established to discuss and resolve problems between and among, for example, exchange carriers, interexchange carriers, and enhanced service providers, concerning communications services.

Established in February 1984, Committee T1 develops technical standards and reports regarding interconnection and interoperability of telecommunications networks at interfaces with end-user systems, carriers, information and enhanced service providers, and customer premises equipment (CPE). Committee T1 has six technical subcommittees that are advised and managed by the T1 Advisory Group (T1AG). Each technical subcommittee develops draft standards and technical reports in its designated areas of expertise. The subcommittees recommend positions on matters under consideration by other national and international standards bodies. Technical subcommittees and their areas of expertise are:

- T1A1: performance and signal processing;
- T1E1: interfaces, power, and protection of networks;
- T1M1: internetwork operations, administration, maintenance, and provisioning;
- T1P1: systems engineering, standards planning, and program management;
- T1S1: services, architectures, and signaling;
- T1X1: digital hierarchy and synchronization.

Membership and participation in Committee T1 are open to all parties with a direct and material interest in the T1 process and activities. Free of dominance by any single interest, T1's policy of open membership and balanced participation safeguards the integrity and efficiency of the standards formulation process. ANSI's due process procedures further ensure fairness. Required procedures include announcing meetings in advance, distributing agendas in advance, adhering to written procedures governing the methods used to develop standards, and giving public notice and opportunity for comment on proposed standards.

The ANSI Board of Standards Review verifies that requirements for due process, consensus, and criteria for standards approval are met on a continual basis.

Scope of Committee T1

Committee T1 (also referred to as "the Committee" or "T1") develops standards and technical reports related to interfaces for U.S. telecommunications networks, some of which are associated with other North American telecommunications networks. T1 develops positions on related subjects under consideration in various international standards bodies. T1 focuses on functions and characteristics associated with interconnection and interoperability of telecommunications networks at interfaces with end-user systems, carriers, and information and enhanced-service providers. These include switching, signaling, transmission, performance, operation, administration, and maintenance aspects. Committee T1 is concerned with procedural matters at points of interconnection, such as maintenance and provisioning methods and documentation, for which standardization would benefit the telecommunications industry.

Responsibilities

The T1 Committee is responsible for:

- Developing proposed American national standards;
- Voting on approval of such proposed standards;
- Maintaining and updating the standards developed by the Committee;
- Interpreting the standards developed by the Committee;
- Adopting and revising Committee policies and procedures;
- Other matters requiring Committee action as provided in the bylaws.

5.2.1.2 T1A1 Performance and Signal Processing

Mission

T1A1 develops and recommends standards and technical reports related to the performance and processing of voice, audio, data, image, and video signals within U.S. telecommunication networks. T1A1 also develops and recommends positions on, and fosters consistency with, standards and related subjects under consideration in other national and international standards bodies.

Scope

The Performance and Signal Processing Technical Subcommittee focuses on two main areas: the performance of networks and services at and between carrier-to-carrier and carrier-to-customer interfaces, with due consideration of end-to-end performance, and the performance of customer systems, and signal processing for the transport and integration of voice, audio, data, image, and video signals with due consideration of:

- Interaction with telecommunications networks;
- The integration of inputs and outputs between information processing systems and telecommunication networks;
- Techniques for assessing the performance and impact of such signal processing on telecommunication networks.

Standards, technical reports, and contributions will be developed in accordance with the following objectives:

- To identify and define performance parameters and levels for speed, accuracy, dependability, and availability of connection establishment, information transfer, and connection disengagement (including survivability parameters)—taking into account characteristics of signal processing systems;

- To define measurement techniques for performance parameters;
- To define methods for characterizing network and signal processing performance for customer applications;
- To identify and develop signal processing algorithms and interface requirements, including coding and compression, interpolation, rate adaptation, echo cancellation, and packetization techniques;
- To take into account interworking of telecommunication networks with customer systems, international networks, other signal processing systems, and new network technologies and services such as asynchronous transfer mode and personal communications.

The T1A1 Committee work is divided into four working groups as defined below.

- T1A1.2: Network Survivability Performance;
- T1A1.3: Digital Network and Service Performance;
- T1A1.5: Multimedia Communications Coding and Performance;
- T1A1.7: Performance and Signal Processing for Voiceband Services.

5.2.1.3   T1E1 Interfaces, Power, and Protection of Networks

Mission

The Interfaces, Power, and Protection of Networks Technical Subcommittee develops and recommends standards and technical reports. The standards and technical reports are related to power systems, electrical and physical protection for the exchange and interexchange carrier networks, and interfaces associated with user access to telecommunications networks.

Scope

The scope of the Interfaces, Power, and Protection for Networks Technical Subcommittee includes, but is not limited to, work on developing standards and technical reports covering the following four areas:

1. Network Interfaces. This work covers the subjects of interfaces (and interface functionality) involving access to the telecommunications (including data communications and integrated services digital) networks by exchange and interexchange carriers, end-users, and enhanced service providers. The work will include the electromagnetic, optical, and

mechanical characteristics of the interfaces, and may include aspects of the physical layer transmission and signaling protocols.

2. Power. This work covers the subjects involving power systems and their user and power interfaces with the DC powered telecommunications equipment.

3. Electrical Protection. This work covers subjects involving electrical protection for the exchange and interexchange carriers networks. It includes grounding systems, electrostatic discharge susceptibility (ESD), electromagnetic interference (EMI), and electromagnetic pulse (EMP).

4. Physical Protection. This work covers areas involving physical protection of the exchange and interexchange networks.

These subjects include but are not limited to temperature, humidity, fire resistance, earthquake resistance, and contamination prevention. There will be close and coordinated working liaison with other T1 technical subcommittees, as well as with external standards setting bodies. The work also includes the development of proposed U.S. contributions to the related work of international standards bodies (e.g., ITU).

5.2.1.4   T1M1 Internetwork Operations, Administration, Maintenance, and Provisioning

Mission

To develop internetwork operations, administration, maintenance and provisioning standards, and technical reports related to interfaces for U.S. telecommunications networks; some of which are associated with other North American telecommunications networks. These standards may apply to planning, engineering, and provisioning of network resources; to operations, maintenance, or administration process; or to requirements and recommendations for support systems and equipment that may be used for these functions. This subcommittee will also develop positions on related subjects under consideration in other domestic and international standards bodies.

Scope

The scope of the Internetwork Operations, Administration, Maintenance, and Provisioning Technical Subcommittee covers standards and reports for internetwork planning and engineering functions such as:

• Traffic routing plans;
• Measurements and forecasts;

- Trunk group planning;
- Circuit and facility ordering;
- Network tones and announcements;
- Location, circuit, equipment identification and other codes; and numbering plans.

The Technical Subcommittee will also consider standards and reports for all aspects of internetwork operations such as network management; circuit and facility installation, line up, restoration, routine maintenance, fault location and repair; contact points for internetwork operations; and service evaluation. The work of the Technical Subcommittee includes standards and reports regarding test equipment and operations support systems together with the required network access and operator interfaces. Further, the Technical Subcommittee is concerned with administrative support functions such as methods for charging, accounting, and billing data. Of necessity, the scope of this work requires a close and coordinated working liaison with other Tl Technical Subcommittees as well as external standards setting bodies.

5.2.1.5 T1P1 Wireless/Mobile Services and Systems

Mission

This Technical Subcommittee (TSC) develops and recommends standards and technical reports related to wireless and/or mobile services and systems, including service descriptions and wireless technologies. This TSC develops and recommends positions on related subjects under consideration in other North American, regional, and international standards bodies.

Scope

The Wireless/Mobile Services and Systems TSC coordinates and develops standards and technical reports primarily relevant to wireless/mobile telecommunications networks in the United States and reviews and prepares contributions on such matters for submission to the appropriate U.S. preparatory body for consideration as ITU contributions or for submission to other domestic and regional standards organizations. The TSC will maintain liaison with other TSCs as well as external forums as appropriate. The TSC will coordinate closely with other standards developing organizations (e.g., TIA, IEEE, ETSI) on wireless issues to ensure that the work programs are complementary.

T1P1 Working Groups

- T1P1.1 International Wireless/Mobile Standards Coordination;

- T1P1.2 Personal Communications Service Descriptions and Network Architectures;
- T1P1.3 Personal Advanced Communications Systems (PACS);
- T1P1.5 PCS 1900;
- T1P1.6 CDMA/TDMA;
- T1P1.7 Wideband-CDMA.

5.2.1.6 T1S1 Services, Architectures, and Signaling

Mission

T1S1 develops and recommends standards and technical reports related to services, architectures, and signaling, in addition to related subjects under consideration in other North American and international standards bodies.

Scope

The Services, Architectures, and Signaling Technical Subcommittee:

- Coordinates and develops standards and technical reports relevant to telecommunications networks in the United States;
- Reviews and prepares contributions on such matters for submission to U.S. ITU-T and U.S. ITU-R Study Groups or other standards organizations;
- Reviews for acceptability or per contra the positions of other countries in related standards development and takes or recommends appropriate actions.

In as much as the development of standards for services, architectures, and signaling may be evolved or influenced by several T1 technical subcommittees, T1S1 will maintain liaison with appropriate technical subcommittees, as well as with standards-setting bodies external to T1.

5.2.1.7 T1X1 Digital Hierarchy and Synchronization

Mission

T1X1 develops and recommends standards and prepares technical reports related to telecommunications network technology pertaining to network synchronization interfaces and hierarchical structures for U.S. telecommunications networks, some of which are associated with other telecommunications networks. T1X1 focuses on those functions and characteristics necessary to define and establish the interconnection of signals comprising network transport. This includes aspects of both asynchronous and synchronous networks. T1X1 also makes recommendations on related subject matter under consideration in various North American and international standards organizations.

Scope

The scope of the work undertaken by the Digital Hierarchy and Synchronization Technical Subcommittee includes the concept, definition, analysis, and documentation of matters pertaining to the interconnection of network transport signals. All theoretical and analytical work necessary to support the documented results is generated or coordinated by the Technical Subcommittee. This requires close liaison with other Committee T1 Technical Subcommittees as well as standards organizations external to Committee T1.

5.2.2 NCITS L3 (Coding of Audio, Picture, Multimedia, and Hypermedia Information)

The L3 Technical Committee on Audio/Picture Coding serves as the U.S. TAG to ISO/IEC JTC 1/SC29, Coding of Audio, Picture, Multimedia, and Hypermedia Information. L3 activities and project development are conducted at the international level for the standardization of coded representation of audio, picture, multimedia, and hypermedia information—and of sets of compression and control functions for use with such information. This information includes: audio information, bi-level and limited bits-per-pixel still pictures; computer graphics images; moving pictures and associated audio, multimedia, and hypermedia information for real-time final form interchange; and audio visual interactive scriptware. The L3 working groups are:

- L3.1—MPEG Development Activity;
- L3.2—Still Image Coding, consisting of JPEG and JBIG;
- L3.3—Multimedia/Hypermedia Information (MHEG).

This technical committee is the U.S. TAG to ISO/IEC JTC 1 SC29 and provides recommendations on U.S. positions to the JTC 1 TAG. Information may be obtained at ncits@itic.nw.dc.us.

5.3 European Telecommunications Standards Institute

The ability for companies to compete in global markets largely depends on their capacity to communicate in a cost effective and reliable manner. In Europe, telecommunications standardization is an important step towards building a harmonized economic market. The European Commission (EC) has set an ambitious pace for achieving a unified single market, and, in addition, the members of the European Free Trade Association and other CEPT countries recognize the benefits of harmonized telecommunications.

The European Telecommunications Standards Institute (ETSI) is a non-profit making organization whose mission is to determine and produce the telecommunications standards that will be used for decades to come. It is an open forum that unites 490 members from 34 countries, representing administrations, network operators, manufacturers, service providers, and users. Any European organization proving an interest in promoting European telecommunications standards has the right to represent that interest in ETSI and thus to directly influence the standards-making process.

It is ETSI members that fix the standards work program in function of market needs. Accordingly, ETSI produces voluntary standards, some of which may go on to be adopted by the EC as the technical base for directives or regulations. Additionally, the fact that the voluntary standards are requested by those who subsequently implement them means that the standards remain practical rather than abstract.

ETSI promotes the world-wide standardization process whenever possible. Its work program is based on, and coordinated with, the activities of international standardization bodies, mainly the ITU-T and the ITU-R.

ETSI consists of a General Assembly, a Board, a Technical Organization, and a Secretariat. The Technical Organization produces and approves technical standards. It encompasses ETSI Projects (EPs), Technical Committees (TCs), and Special Committees. More than 3,500 experts are at present working for ETSI in over 200 groups.

The central Secretariat of ETSI is located in Sophia Antipolis, a high-tech research park in Southern France. It comprises about 100 staff members. At present, there are about 30 Specialist Task Forces (STFs) with around 100 experts. Up to now, over 2,800 ETSI deliverables have been published.

The technical organization of ETSI comprises the three types of Technical Bodies, designated either as an ETSI project, a Technical Committee, or an ETSI Partnership Project. They provide the forum for technical discussion and have as their main tasks the preparation of work programs and ETSI deliverables. A Technical Body is the primary decision making center for all matters that fall within its Terms of Reference.

A Technical Body may establish working groups if required to undertake specific areas of work. A Technical Body may also use specialist task forces to carry out well-defined tasks within specified periods of time in order to speed up urgent items.

6

Multimedia Communications via N-ISDN (H.320)

6.1 History

The ITU is a part of the United Nations, and its purpose is to develop formal "Recommendations" to ensure worldwide communications are accomplished efficiently and effectively. In 1984, the first recommendations for a video teleconferencing codec (H.120 and H.130) were established. These recommendations were defined specifically for the European region (625 lines, 2.048-Mbps primary rate) and for interconnection between Europe and other regions. Since no recommendation existed for non-European regions, it lacked true international scope, and in 1984 the ITU established a Specialists Group on Coding for Visual Telephony to develop a truly international recommendation. The ITU established two objectives for the Specialists Group: to develop a recommendation for a video codec for teleconferencing application operating at the bit rates of N $\times$ 384 Kbps (N = 1 through 5), and to begin the standardization process for a video codec for teleconferencing and video telephone applications operating at bit rates of M $\times$ 64 Kbps (M-1, 2).

The Specialists Group, chaired by S. Okubo from Japan, met 17 times between 1984 through 1989. At the September meeting in 1988, it was determined that the compression algorithm chosen for N $\times$ 384 Kbps was sufficiently flexible that it could be extended, with good performance, down to 64 Kbps. At that time, the Specialists Group shifted their focus to develop a single recommendation to code at all bit rates from 64 Kbps to 2 Mbps, that is, to code at rates of p $\times$ 64 Kbps, where the key values of p are 1, 2, 6, 24, and 30.

In 1989, a number of organizations in Europe, the United States, and Japan developed flexible codec systems to meet a preliminary specification of the standard. Various systems were interconnected in the laboratory and by long-distance communication channels to validate the recommendation. These tests were highly successful and encouraging. The ITU formally approved the H.320 series of standards in December 1990.

6.2   H.320: The VTC System Standard

H.320 is the overview ITU recommendation that specifies a multimedia terminal for transmission over the N-ISDN network. The entire set of five standards that fully defines the H.320 terminal is listed in Table 6.1.

Since the five H.320 recommendations were finalized by the ITU in December 1990, the standard is extremely mature and forms the cornerstone of the videoconferencing business.

Figure 6.1 illustrates the interrelationship of the five H.320 standards and shows the connections of the H.320 terminal with external components. Highlights of key aspects of the H.320 multimedia terminal follow.

One function of the H.320 recommendation is to define the phases of establishing a visual telephone call as listed below.

- Phase A: Call set-up, out-band signaling;
- Phase B1: Mode initialization on initial channel;
- Phase CA: Call set-up of additional channel(s), if relevant;
- Phase CB1: Initialization on additional channel(s);
- Phase B2 (or CB2): Establishment of common parameters;
- Phase C: Visual telephone communication;
- Phase D: Termination phase;
- Phase E: Call release.

Another function of Recommendation H.320 is the definition of 16 types of visual telephone terminals and their modes of operation (see Tables 6.2 and 6.3, respectively).

6.3   H.261: Video Coding Standard

If the standard TV signal were to be encoded using conventional 8-bit PCM, a bit rate of approximately 90 Mbps would be required for transmission. Video

Table 6.1
H.320 Recommendation Series

Designation	Title	Purpose
H.320	Narrowband visual telephone systems and terminal equipment	This standard defines the interrelationship of all of the five H.320 recommendations.
H.261	Video codec for audiovisual services at P × 64 Kbps	This recommendation specifies the video coding algorithm, the picture format, and forward error-correction techniques for the audiovisual terminal.
H.221	Frame structure for a 64- to 1,920-Kbps channel in audiovisual teleservices.	The purpose of this recommendation is to define a frame structure for audiovisual teleservices in single or multiple B or HO channels or a single H11 or H12 channel.
H.242	System for establishing communication between audiovisual terminals using digital channels up to 2 Mbps	Recommendation H.242 defines the detailed "handshake" protocol and procedures that are employed by H.320 terminals in the preliminary phases of a call.
H.230	Frame synchronous control and indication (C&I) signals for audiovisual systems	Recommendation H.230 has two primary elements: (1) it defines the C&I symbols related to video, audio, maintenance, and multipoint; (2) it contains a table of bit rate allocation signal (BAS) escape codes which clarifies the circumstances under which some C&I functions are mandatory and others optional.

compression technology is used to reduce this bit rate to the primary rates (1.544 and 2.048 Mbps), fractional primary rates (e.g., 384 Kbps), and basic rates (64 Kbps and multiples), which are employed for economical transmission. The compression function is performed by a video codec (COder, DECoder), and H.261 is the ITU recommendation for the video codec for teleconferencing.

Figure 6.2 is a functional block diagram of the video codec as defined in Recommendation H.261. The heart of the system is the source coder, which

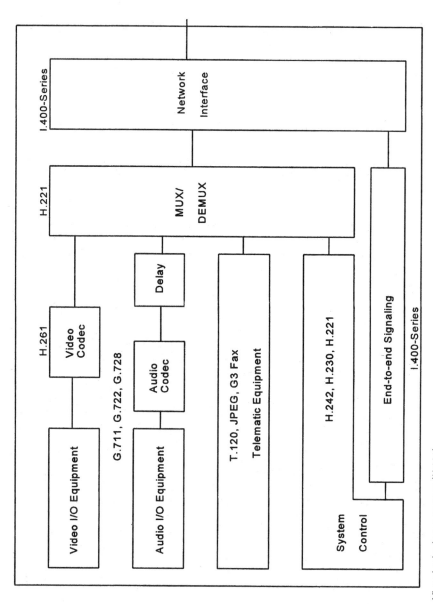

Figure 6.1 Visual telephone system (H.320).

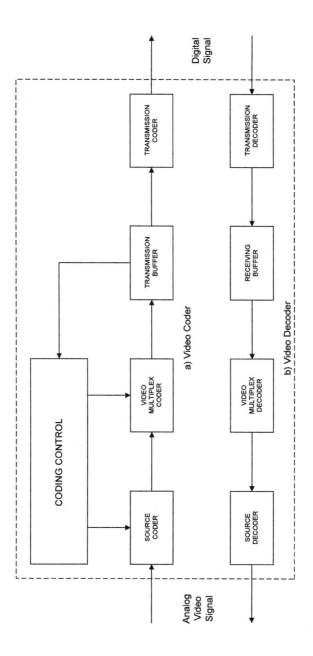

Figure 6.2 Block diagram of the video codec.

Table 6.2
Communication Modes of Visual Telephone

Visual Telephone Mode		Channel Rate (Kbps)	ISDN Channel (NOTE 2)	ISDN Interface Basic	ISDN Interface Primary Rate	Coding Audio	Coding Video
a	a$_0$	64	B			Rec. G.711 (NOTE 4)	Rec. H.261 (NOTE 6)
	a$_1$					Rec. G.728	
b	b$_1$	128	2B			Rec. G.711	
	b$_2$					Rec. G.722	
	b$_3$					Rec. G.728	
q (NOTE 3)	q$_1$	$n \times 64$	nB			Rec. G.711	
	q$_2$					Rec. G.722	
	q$_3$					Rec. G.728	
g		384	H$_0$	Not applicable	Applicable	(NOTE 5)	Rec. H.261
h		768	2H$_0$				
i		1,152	3H$_0$				
j		1,536	4H$_0$				
k		1,536	H$_{11}$				
l		1,920	5H$_0$				
m		1,920	H$_{12}$				

Table 6.2 *(continued)*

NOTE 1—(Audio coding of mode b_3) In addition to G.728, higher quality audio coding such as H.200/AV.253 may be used for this mode.

NOTE 2—For multiple channels of B/H_0, all channels are synchronized at the terminal according to 2.7/H.221.

NOTE 3—$q = c/d/e/f$ corresponds to $n = 3/4/5/6$, respectively. This mode is applicable to the ISDN basic interface if multiple basic accesses are used.

NOTE 4—If a visual telephone interworks with a wideband speech terminal, G.722 audio may be used instead of G.711 audio.

NOTE 5—Modes (G.711 and G.728 audio) other than this recommended mode may be invoked by H.242 procedures.

NOTE 6—If two terminals connect at this rate and run G.711 and both have video capability, H.261 may be used. It should be noted, however, that the video performance is limited due to the very low bit rate available for this purpose.

Table 6.3
Visual Telephone Terminal Types

Mode		Type X (NOTE 2)									Type Y (NOTE 3)					Type Z
Transfer rate	Audio coding	a	b_1	$b_{2/3}$	b_4	b_5	q_1	$q_{2/3}$	q_4	q_5	1	2	3	4	5	
a_0 B	G.711 (NOTE 4)	X	X	X	X	X	X	X	X	X						
a_1 B	G.728	X	X	X			X	X								
b_1 2B	G.711		X	X	X	X	X	X	X	X						
b_2 2B	G.722			X		X		X		X						
b_3 2B	G.728		X	X			X	X								
q_1 nB	G.711 (NOTE 5)				X	X	X	X	X	X						
q_2 nB	G.722 (NOTE 5)							X	X	X						
q_3 nB	G.728 (NOTE 5)						X	X								
g H_0	G.722										X	X	X	X	X	
h $2H_0$	G.722											X	X	X	X	
i $3H_0$	G.722												X	X	X	
j $4H_0$	G.722													X	X	
k H_{11}	G.722															X

Table 6.3 *(continued)*

Mode		Type X (NOTE 2)									Type Y (NOTE 3)					Type Z
Transfer rate	Audio coding	a	b_1	$b_{2/3}$	b_4	b_5	q_1	$q_{2/3}$	q_4	q_5	1	2	3	4	5	
l 5H$_0$	G.722															
m H$_{12}$	G.722														X	X

NOTE 1—"X" means the terminal of the given type is able to work in the given mode.

NOTE 2—Types XB$_4$ and Xb$_5$ are defined to take into account that G.728 had not yet been recommended when this recommendation was first established.

NOTE 3—Terminal of this type must conform to § 3.3.2.2.

NOTE 4—If a visual telephone interworks with a wideband speech terminal, G.722 audio may be used instead of G.711 audio.

NOTE 5—$q = c/d/e/f$ corresponds to $n = 3/4/5/6$, respectively. Since transfer rates of multiple B are defined hierarchically, Type X$_{f_1}$, for example, supports all of (a_1, b_1, c_1, d_1, e_1, f_1) and (b_3, c_3, d_3, e_3, f_3) modes.

compresses the incoming video signal by reducing redundancy inherent in the TV signal. The multiplexer combines the compressed data with various side information, which indicates alternative modes of operation. A transmission buffer is employed to smooth the varying bit rate from the source encoder to adapt it for the fixed bit-rate communication channel. A transmission coder includes functions such as forward error-control to prepare the signal for the data link.

One of the most critical design issues for a video codec is the control of the basic buffer located between the coder and the channel. The buffer fullness is continually monitored. If the buffer is threatening to overflow, the coding operation is quickly modified by reducing the transmitted frame rate, reducing the number of transmitted coefficients, or coding the coefficients more coarsely. Conversely, if the buffer is threatening to underflow, the coding parameters are modified to transmit more information. The strategy to control the buffer fullness by modification of coding parameters is not standardized. Consequently it serves as a discriminator between codec manufacturers.

One of the most challenging problems to be solved by the codec was the reconciliation of the incompatibility between European TV standards (for example, PAL and SECAM) and that used in most other areas of the world, NTSC. Phase alteration by line (PAL) and sequential color avec memoir (SECAM) employ 625 lines and a 50-Hz field rate, while NTSC has 525 lines and a 60-Hz field rate. This conflict was resolved by adopting a spatial resolution standard based on the European standard and a temporal resolution based on NTSC (see Table 6.4).

The QCIF format, which employs half the CIF spatial resolution in both horizontal and vertical directions, is the mandatory H.261 format; full CIF is optional. In general, QCIF is used for videophone applications where head-and-shoulders pictures are sent from desk to desk. Conversely, the full CIF format is usually used for teleconferencing where several people must be viewed in a conference room.

Table 6.4
H.261 Picture Formats

Parameter	CIF	QCIF
Coded pictures per second	29.97	(or integral submultiples)
Coded luminance pixels per line	352	176
Coded luminance lines per picture	288	144
Coded color pixels per line	176	88
Coded color lines per picture	144	72

Figure 6.3 is a functional block diagram outlining the H.261 source coder. Interframe prediction is first carried out in the pixel domain. The prediction errors are encoded by the discrete cosine transform using blocks of 8 × 8 pels. The transform coefficients are next quantized and fed to the multiplexer. Motion compensation is included in the prediction on an optional basis.

6.3.1 Picture Structure

In the encoding process, each picture is subdivided into groups of blocks (GOB). As shown in Figure 6.4, the CIF picture is divided into 12 GOBs while QCIF has only three GOBs. From the GOB level down, the structures of CIF and QCIF are identical. A header at the beginning of the GOB permits changing the coding accuracy and resynchronization in case errors are encountered.

Each GOB is further divided into 33 macroblocks, as shown in Figure 6.5. The macroblock header defines the location of the macroblock within the GOB, the type of coding to be performed, possible motion vectors, and which blocks within the macroblock will actually be coded. There are two basic types of coding. In intra coding, coding is performed without reference to previous pictures. This mode is relatively rare but is required for forced updating, and every macroblock must occasionally be intra coded to control the accumulation of inverse transform mismatch errors. The more common coding type is inter, in which only the difference between the previous and the current pictures is coded. Of course, for picture areas without motion, the macroblock does not have to be coded at all.

Each macroblock is further divided into six blocks, as shown in Figure 6.6. Four of the blocks represent the luminance, or brightness, while the other two represent the red and blue color differences for the entire macroblock. Each block is 8 × 8 pixels, so it can be seen that the color resolution is half of the luminance resolution in both dimensions.

6.3.2 Example of Block Coding

Figure 6.7 shows a simple example of how each 8 × 8 block is coded. In this case, intra coding is used, but the principle is the same for inter coding. Figure 6.7(a) shows the original block to be coded. Without compression, this would take 8 bits to code each of the 64 pixels, or a total of 512 bits. First, the block is transformed, using the two-dimensional DCT, giving the coefficients of Figure 6.7(b). Note that most of the energy is concentrated into the upper left-hand corner of the coefficient matrix. Next, the coefficients of Figure 6.7(b) are quantized with a step size of 6. (The first term, DC, always uses a step size of 8.) This produces the values of Figure 6.7(c), which are much smaller in magnitude than

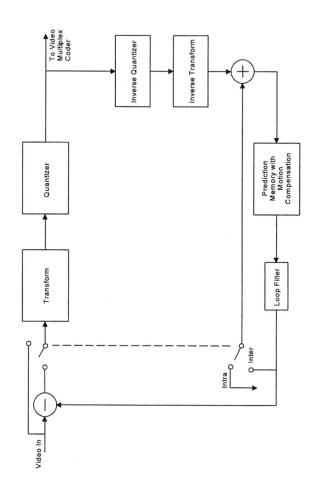

Figure 6.3 Source coder.

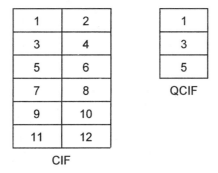

Figure 6.4 Arrangement of GOBs in a picture.

the original coefficients, and most of the coefficients become zero. The larger the step size, the smaller the values produced, resulting in more compression.

The coefficients are then reordered, using the zigzag scanning order of Figure 6.8. All zero coefficients are replaced with a count of the number of zeros before each nonzero coefficient (RUN). Each combination of RUN and VALUE produces a variable length code that is sent to the decoder. The last nonzero VALUE is followed by an end of block (EoB) code. The total number of bits used to describe the block is 25, a compression of 20:1.

At the decoder (and at the coder to produce the prediction picture), the step size and VALUEs are used to reconstruct the inverse quantized coefficients, which, as shown in Figure 6.7(e), are similar to, but not exactly equal to, the original coefficients. When these coefficients are inverse transformed, the result of Figure 6.7(f) is obtained. Note that the differences between this block and the original block are quite small.

6.3.3 Motion Compensation

The operation of motion compensation is shown in Figure 6.9. Block A is a block in the current picture that is to be coded. Block B is the block at the same position as A but in the picture that was previously stored in both coder and decoder. Because of image motion, block A more closely resembles the pixel

1	2	3	4	5	6	7	8	9	10	11
12	13	14	15	16	17	18	19	20	21	22
23	24	25	26	27	28	29	30	31	32	33

Figure 6.5 Arrangement of macroblocks in a GOB.

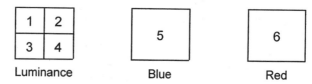

Figure 6.6 Arrangement of blocks in a macroblock.

75	76	77	78	79	80	81	82
77	78	79	80	81	82	83	84
79	80	81	82	83	84	85	86
81	82	83	84	85	86	87	88
83	84	85	86	87	88	89	90
85	86	87	88	89	90	91	92
87	88	89	90	91	92	93	94
89	90	91	92	93	94	95	96

a) ORIGINAL BLOCK (8x8x8 = 512 BITS)

76	76	77	79	80	81	82	83
77	77	78	80	81	82	83	84
79	79	80	81	83	84	85	86
81	82	83	84	85	87	88	88
84	84	85	87	88	89	90	91
86	87	88	89	91	92	93	93
88	89	90	91	92	94	95	95
89	90	91	92	93	95	96	96

f) RECONSTITUTED BLOCK

684	-19	-1	-2	0	-1	0	-1
-37	0	-1	0	0	0	0	-1
0	0	0	0	0	0	0	0
-4	-1	-1	-1	-1	0	-1	-1
0	0	0	0	0	0	0	0
-2	0	0	-1	0	-1	0	-1
0	0	0	0	-1	-1	-1	-1
-1	-1	-1	0	-1	0	-1	0

b) TRANSFORMED BLOCK COEFFICIENTS

688	-21	0	0	0	0	0	0
-39	0	0	0	0	0	0	0
0	0	0	0	0	0	0	0
0	0	0	0	0	0	0	0
0	0	0	0	0	0	0	0
0	0	0	0	0	0	0	0
0	0	0	0	0	0	0	0
0	0	0	0	0	0	0	0

e) INVERSE QUANTIZED COEFFICIENTS

86	-3	0	0	0	0	0	0
-6	0	0	0	0	0	0	0
0	0	0	0	0	0	0	0
0	0	0	0	0	0	0	0
0	0	0	0	0	0	0	0
0	0	0	0	0	0	0	0
0	0	0	0	0	0	0	0
0	0	0	0	0	0	0	0

c) QUANTIZED COEFFICIENT LEVELS

RUN LEVEL CODE

0	86	01010110
0	-3	001011
0	-6	001000011
	EOB	10

TOTAL CODE LENGTH = 25

d)   COEFFICIENTS IN ZIG-ZAG ORDER AND VARIABLE LENGTH CODED

Figure 6.7 Sample intra block coding.

1	2	6	7	15	16	28	29
3	5	8	14	17	27	30	43
4	9	13	18	26	31	42	44
10	12	19	25	32	41	45	54
11	20	24	33	40	46	53	55
21	23	34	39	47	52	56	61
22	35	38	48	51	57	60	62
36	37	49	50	58	59	63	64

Figure 6.8   Scanning order in a block.

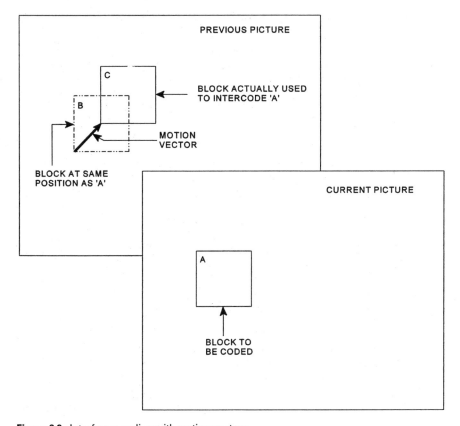

Figure 6.9   Interframe coding with motion vectors.

data from block C than that from block B. The displacement of block C from block B, measured in pixels in the x- and y-directions, is the motion vector. The pixel-by-pixel difference between blocks A and C is transformed and coded. The motion vector and code data are transmitted to the decoder, where the inverse transformed block data is added to the data in block C pointed to by the motion vector and placed in the block A position.

The use of motion vectors is optional in the coder, where the calculation of the optimum motion vectors is complex, but required in the decoder, where the reconstruction of the motion is relatively simple.

Figure 6.10 is an illustration of the decision tree which is used to encode macroblocks. The last step in the coding tree provides for the option to change the quantization level used for the macroblock.

The H.261 standard does not define all aspects of image coding and decoding. Rather, it is just an interoperability specification, guaranteeing that any codec manufactured according to the standard will be able to communicate with each other. This still allows considerable freedom for manufacturers to offer better performance, and new developments can be incorporated. (This is in contrast with the G.722 audio standard, where the encoding algorithm is rather precisely defined.) For example, the encoder strategy is not defined. Which blocks will be encoded, with what type of code, and with what accuracy is under control of the designer. While there is less freedom for the decoder, post-processing, such as error concealment, filtering, or interpolation of the image, is under control of the designer.

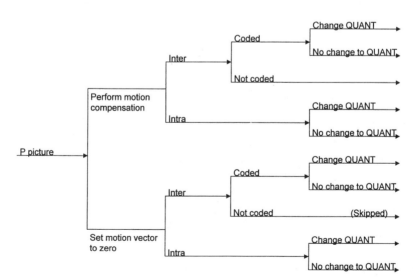

Figure 6.10 Decision tree for coding macroblocks.

Furthermore, the H.242 Recommendation permits two codecs to negotiate to an altogether different proprietary algorithm that they both incorporate, with the H.261 algorithm becoming a fall-back mode for codecs of different manufacture.

It should be noted that the newly developed H.263 video coding standard (see Section 8.4) has been added to the H.320 as an option resulting in improved video coding over the N-ISDN.

6.4   H.221: Frame Structure for a 64- to 20-Kbps Channel in Audiovisual Teleservices

The purpose of this recommendation is to define a frame structure for audiovisual teleservices in single or multiple B or HO channels, or a single H11 or H12 channel that makes the best use of the characteristics and properties of the audio and video encoding algorithms, the transmission frame structure, and the existing ITU recommendations. It offers several advantages, as follows.

• It is simple, economic, and flexible. It may be implemented on a simple microprocessor using well-known hardware principles.

• It is a synchronous procedure. The exact time of a configuration change is the same in the transmitter and the receiver. Configurations can be changed at 20-ms intervals.

• It needs no return link for audiovisual signal transmission, since a configuration is signaled by repeatedly transmitted codewords.

• It is very secure in case of transmission errors, since the code controlling the multiplex is protected by a double error correcting code.

• It allows the synchronization of multiple 64- or 384-Kbps connections and the control of the multiplexing of audio, video, data, and other signals within the synchronized multiconnection structure in the case of multimedia services such as videoconference.

This recommendation provides for dynamically subdividing an overall transmission channel of 64 to 1,920 Kbps into lower rates suitable for audio, video, data, and telematic purposes. The overall transmission channel is derived by synchronizing and ordering transmissions over from 1B to 6B connections, from 1HO to 5HO connections, or an H11 or H12 connection.

A single 64-Kbps channel is structured into octets transmitted at 8 kHz. Each bit position of the octets may be regarded as a subchannel of 8 Kbps. The eighth subchannel is called the service channel (SC), containing the two critical parts.

- *Frame alignment signal* (FAS): This 8-bit code is used to frame the 80 octets of information in a B channel.
- *Bit rate allocation signal* (BAS): This 8-bit code describes the capability of a terminal to structure the capacity of the channel or synchronized multiple channels in various ways and to command a receiver to demultiplex and make use of the constituent signals in such structures. This signal is also used for controls and indications.

The video bitstream is carried in frames of data as shown in Figure 6.11. Each frame corresponds to a 64-Kbps B channel in ISDN. Two frames are shown, one for the audio portion of the conference, and the other for the video portion. In each, there is an 8-bit FAS that permits synchronization of the frame and low-speed signaling of communication overhead. There is also an 8-bit BAS that defines how the H.221 channels and subchannels are divided and what type of service is used on each section. For example, one BAS code is used for standard video to Recommendation H.261, while another might indicate that two B channels are allocated to this service. The BAS codes can change from frame to frame to indicate complex protocols or changes of mode.

Each frame of 640 bits is transmitted in 10 ms, giving an overall rate of 64 Kbps. However, the FAS and BAS use 16 of the 640 bits, so the net rate available for video is only 62.4 Kbps for a single B channel. The order of transmission is left to right across each row and then the row below. Higher bit rates can be obtained by using multiple B channels (up to 2 for ISDN basic access, up to 30 for primary access).

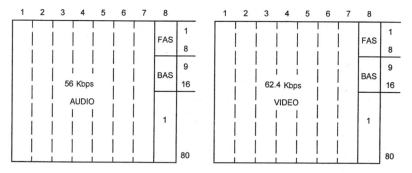

Figure 6.11 H.221 frames.

6.5 H.242: System for Establishing Communication Between Audiovisual Terminals Using Digital Channels Up to 2 Mbps

Recommendation H.242 defines the detailed *handshake* protocol and procedures that are employed by H.320 terminals in the preliminary phases of a call. Major topics covered in this recommendation are:

- Basic sequences for in-channel procedures;
- Mode initialization, dynamic mode switching, and mode 0 forcing recovery from fault conditions;
- Network consideration: call connections, disconnection, and call transfer;
- Procedures for activation and deactivation of data channels;
- Procedures for operation of terminals in restricted networks.

6.6 H.230: Frame Synchronous Control and Indication Signals for Audiovisual Systems

Digital audiovisual services are provided by a transmission system in which the relevant signals are multiplexed onto a digital path. In addition to the audio, video, user data, and telematic information, these signals include information for the proper functioning of the system. The additional information has been named control and indication (C&I) to reflect the fact that while some bits are genuinely for control, causing a state change somewhere else in the system, others provide for indications to the users as to the functioning of the system.

Recommendation H.230 has two primary elements. First, it defines the C&I symbols related to video, audio, maintenance, and multipoint. Second, it contains a table of BAS escape codes that clarifies the circumstances under which some C&I functions are mandatory and others optional.

6.7 Audio Coding

The BAS codes of H.221 are used to signal a wide range of possible audio coding modes. The most prominent modes define existing ITU Recommendations G.711, G.722, and G.728 (see Table 6.5). Recommendation G.711 (pulse code modulation of voice frequencies) is used for narrowband speech since it samples only at 8,000 samples/s and encodes to 8 bits/sample for a transmission rate of 64 Kbps.

Table 6.5
ITU Audio Coding Standards

ITU Recommendation	Audio Bandwidth (kHz)	Bit Rate (Kbps)	Coding Algorithm
G.711	3	64	PCM
G.722	7	64, 56, 48	Subband adaptive differential PCM
G.728	3	16	CELP

Recommendation G.722 (7-kHz audio coding with 64 Kbps) describes the characteristics of an audio (50- to 7,000-Hz) coding system that may be used for a variety of higher quality speech applications, The coding system uses subband adaptive differential pulse code modulation (SB-ADPCM) within a bit rate of 64 Kbps. In the SB-ADPCM technique used, the frequency band is split into two subbands (higher and lower) and the signals in each subband are encoded using ADPCM. The system has three basic modes of operation corresponding to the used bit rates of 7-kHz audio coding: 64, 56, and 48 Kbps.

The G.711 and G.722 standards have been used extensively in P × 64 systems operating at high bit rates (for example, 1.544 Mbps and 768 Kbps) where the 64 Kbps for audio is a relatively small percentage of the transmission bit rate. However, in recent years, the channel bit rate has been typically reduced to 112 Kbps, where 64-Kbps audio would require too large a fraction of the channel capacity. The recent G.728 standard operating at 16 Kbps is a major development that alleviates this crowding. Two audio coding standards (G.723.1 and G.729) that have been developed very recently for low bit-rate operation (6.8 and 8 Kbps, respectively) have been added to H.320 as potential optional speech coding techniques.

6.8 Data Channel

The H.320 standards provide for the transmission of data along with video and audio information. Low-speed data (LSD) and high-speed data (HSD) channels are specified. See Figure 6.12. These operate up to 64 Kbps and at multiples of 64 Kbps, respectively. For these channels to be useful it is necessary to develop communication protocols and applications to define data flow. Figure 6.12 shows that the T.126 (still image/annotation) and T.127 (binary file transfer) application protocols have been developed for use over the T.120 communication protocol stack and the multilayer protocol (MLP) data channel as defined

APPLICATIONS		T.121 - Audiograph Conferencing T.126 - Still Image T.res - Reservations T.avc - Audio & Video Control T.bwc - Bandwidth Control T.127 - Binary File Transfer	H.280 - Far End Camera Control (Simplex, low delay)
COMMUNICATION PROTOCOL		T.120 Series - T.123 Protocol Stack - T.122 Multipoint Comm. Svc. - T.124 Conference Control - T.125 Multipoint Protocol	H.224 - Data Link Layer (DLL)
DATA CHANNELS	Low Speed (<64 Kbps)	MLP (4, 6.4, VAR)	LSD (Low Speed Data)
	High Speed (Multiples of 64 Kbps)	H-MLP	HSD (High Speed Data)

Figure 6.12 Data transmission in H.320.

by the ITU Study Group 8. The H.280 (far-end camera control) application protocol has also been developed for use over the H.224 (data link layer) communications protocol.

6.9   H.320 Range of Capabilities

Table 6.6 is a summary of the capabilities of the H.320 standard system, ranging from the mandatory minimum parameters up to the optional maximum values. The only parameter which has not been previously discussed is the range of search for motion compensation. The H.261 standard specifies the *maximum* length of a motion vector used in motion compensation to be plus or minus 15 pixels. This means that, if the encoder was designed to take full advantage of the motion compensation potential, a range of 30×30 pixels must be searched to find the best match for the block to be transmitted. It should be noted that this search process consumes a very large fraction of encoder resources. Some manufacturers have simplified their encoder by reducing the range of the search process. In other words, they have chosen to implement motion compensation, but are unable to search beyond a reduced range of 16×16 pixels, for example. This is a classic example of the tradeoff between cost and performance.

6.10   Multipoint

In 1992, the ITU approved Recommendations H.231, Multipoint Control Unit for Audiovisual Systems Using Digital Channels Up to 2 Mbps, and

Table 6.6
H.320 Range of Capabilities

	Mandatory Minimum	Optional Maximum
Resolution	QCIF 176 × 144	CIF 352 × 288
Frame rate	Up to 7.5 fps	Up to 30 fps
Transmission bit rate	56/64 Kbps	2,048 Kbps
Motion compensation		
- Encoder	No	Yes (30 × 30 pixel search)
- Decoder	Yes	Yes
Pre/post processing	None	Extensive (error concealment)
Audio coding	G.711	G.722, G.728, G.723.1
T.120 data capability	None	Extensive capability

H.243, System for Establishing Communication Between Three or More Audiovisual Terminals Using Digital Channels Up to 2 Mbps.

As the name implies, H.231 defines a multipoint control unit (MCU) that serves as a bridge in multipoint connections. Recommendation H.243 defines the communication protocol between an H.320 terminal and an H.231 MCU. The MCU enables three or more H.320 terminals to participate in an audiovisual conference. Two or more MCUs can be cascaded as illustrated in Figure 6.13. The MCU provides audio mixing and a video switching capability that provides for video teleconferencing.

In multipoint systems, the user may be presented with continuous presence (sometimes called "Hollywood squares") or switched displays. Continuous presence involves spatially multiplexing the selected images into a single image

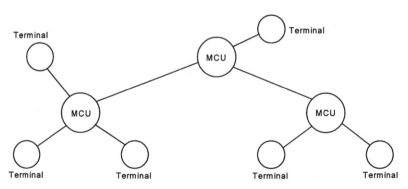

Figure 6.13 Typical multipoint configuration.

in a split screen format. In the switched mode, all participants view the signal originating from one of the other conference participants. There are three primary ways to control the switching process to determine which of the other users is viewed: voice-activated switching, user selection, and chair control. In the case of voice activation, the *speaker's* image is automatically presented to all conferees. In the case of chair control, the conference chairman has control over which participant is viewed by all users.

6.11 Privacy

The ITU has developed two recommendations to provide for the privacy of transmission between H.320 audiovisual terminals. A privacy system consists of the confidentiality mechanism or encryption process for the data and a key management subsystem as shown in Figure 6.14.

H.233 describes the confidentiality part of a privacy system suitable for H.320 videoconferencing terminals. Although an encryption algorithm is required for such a privacy system, the specification of such an algorithm is not included. The recommendations refer to a number of possible algorithms (for example, DES, FEAL, and BCRYPT), and more are being added in two phases. Phase I for point-to-point encryption was completed in 1993, while multipoint encryption may be defined later.

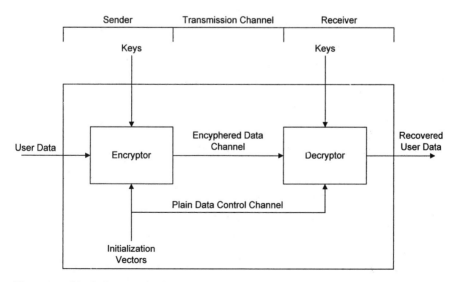

Figure 6.14 Block diagram of a link encryptor.

Recommendation H.234 describes the authentication and key management methods for a privacy system suitable for H.230 terminals. Privacy is achieved by using secret keys. The keys are loaded into the confidentiality part of the privacy system and control the way in which the transmitted data are encrypted and decrypted. If a third party gains access to the keys being used, then the privacy system is no longer secure.

6.12 Procedures for Establishing Communication Between Two Multiprotocol Audiovisual Terminals Using Digital Channels (V.140)

In January 1998, the ITU approved Recommendation V.140, the purpose of which is to automatically negotiate preferred modes of operation between multimedia terminals having multiple protocol capabilities. This recommendation defines automatic mode negotiation and selection for multiprotocol audiovisual terminals connected to digital networks such as ISDN. The procedures in this standard are intended to avoid interference with those of pre-existing recommendations. The procedures in this recommendation automatically determine network connectivity and bit alignment between terminals. Also, they enable prompt and accurate negotiation of a common mode of operation when one or both terminals support multiple protocols for audiovisual communication. For example, a terminal might support H.320, H.324 over voice-band modem, and H.323 over ISDN; in this case, the procedures in this recommendation would be used to negotiate a common protocol (e.g., H.323). Once a mode is selected, further negotiations as recommended for that mode may use information derived from V.140 negotiations, if applicable.

The procedures of V.140 can also be used to provide an optional initial voice telephony mode before proceeding to multimedia telephony, and to switch from one multimedia telephony mode to another, or back into voice telephony mode.

The means by which digital channels are established between terminals are outside the scope of this recommendation (see Recommendation H.220/AV.420). Information regarding the nature of the endpoint terminals available from D-channel signaling may be useful in further accelerating V.140 negotiations.

The procedures of this recommendation concern only the flow of signals along the fixed digital paths used for the transport of audiovisual content during the call, at integer multiples of 64 Kbps (or 56 Kbps in certain networks).

Two primary benefits from implementing V.140 are:

1. Improved reliability in completing calls and successfully establishing multimedia communications, since peculiar network characteristics

(such as "restricted" networks) that interfere with call establishment and protocol negotiations are automatically handled by V.140 procedures.

2. For terminals that support multiple communication modes, an automatic means of selecting modes.

Principal features of V.140 include the ability to find bit alignment, detect remote network type, and perform in-band testing of channel characteristics; a flexible and extensible capability exchange and mode selection facility is also built-in.

V.140 procedures apply to every channel of a multichannel call and begin following establishment of the end-to-end digital connection and before any multimedia or other communication protocols are initiated. The procedures are divided into three phases.

- *Phase 1:* Send/search for V.140 signature (note that V.8/V.8bis, voice, and H.221 FAS, or any subset of these may be simultaneously transmitted). If such signature is detected, proceed to:

- *Phase 2:* Characterize the digital connection (64 versus 56 Kbps, detect octet/septet alignment), and diagnose any odd characteristics of the network (i.e., it could be restricted, that is, it transfers only 7 or 8 bits to the far-end). Once this is complete:

- *Phase 3:* Exchange mode capabilities (similar to V.8bis) and select desired operation mode. Modes can include voice communications, multimedia communications, and channel aggregation protocols, but capabilities are kept simple, given that the purpose of V.140 is only to select a particular protocol, not to determine all of the parameters related to that protocol (which can typically be determined using the protocol itself).

The Phase 1 procedures have been designed to allow simultaneously signaling of other protocols, such as H.320, to minimize the amount of time lost in beginning communications if a terminal that implements V.140 finds itself communicating with a terminal that does not.

Following Phase 3, a terminal can immediately begin procedures associated with the selected operation mode.

6.13 Broadcast Version of H.320 (H.331)

Conventional H.320 systems require a two-way communication channel to perform interactive functions such as mode negotiation. In a broadcast application, the return communication channel is not available. The ITU has developed a version of H.320 (H.331) which does not require a return channel for broadcast operation.

6.14 Narrowband ISDN

The digital pipe between the central office and the ISDN subscriber will be used to carry a number of communication channels. The capacity of the pipe, and therefore the number of channels carried, may vary from user to user. The transmission structure of any access link will be constructed from the following types of channels:

D channel (basic access):	16 Kbps
D channel (primary access):	64 Kbps
B channel:	64 Kbps
H0 channel:	384 Kbps
H11 channel:	1.536 Mbps
H.12 channel:	1.920 Mbps

The B channel is a user channel that can be used to carry digital data, PCM-encoded digital voice, or a mixture of lower rate traffic including digital data and digitized voice encoded at a fraction of 64 Kbps. In the case of mixed traffic, all traffic of the B channel must be destined for the same endpoint; that is, the elemental unit of circuit switching is the B channel.

The D channel serves two main purposes. First, it carries common-channel signaling information to control circuit-switched calls on associated B channels at the user interface. In addition, the D channel may be used for packet switching or low-speed (for example, 100 bps) telemetry at times when no signaling information is waiting.

H channels are provided for user information at higher bit rates. The user may use such a channel as a high-speed trunk or subdivide the channel according to the user's own TDM scheme. Examples of applications include fast facsimile, video, high-speed data, high-quality audio, and multiplexed information streams at lower data rates.

These channel types are grouped into transmission structures that are offered as a package to the user. The best-defined structures at this time are the basic channel structure "basic access," and the primary channel structure "primary access."

6.14.1 Basic Access

Basic access consists of two full-duplex 64-Kbps B channels and a full-duplex 16-Kbps D channel. The total bit rate, by simple arithmetic, is 144 Kbps. However, framing, synchronization, and other overhead bits bring the total bit rate on a basic access link to 192 Kbps. The basic service is intended to meet the needs of most individual users, including residential subscribers, and small offices. It allows the simultaneous use of voice and other applications, such as videoconferencing. Most existing two-wire local loops can support this interface. The basic access (BRI) is probably the most common connection for videoconferencing today. The useful bit rate is 128 Kbps for unrestricted networks and 112 Kbps for restricted networks.

6.14.2 Primary Access

Primary access is intended for users with greater capacity requirements, such as offices with a digital PBX or a LAN. Because of differences in the digital transmission hierarchies used in different countries, it was not possible to get agreement on a single data rate. The United States, Canada, and Japan make use of a transmission structure based on 1.544 Mbps; this corresponds to the T1 transmission facility of AT&T. In Europe, 2.048 Mbps is the standard rate. Both of these data rates are provided as a primary interface service. Typically, the channel structure for the 1.544-Mbps rate will be 23 B channels plus one 64-Kbps D channel and for the 2.048-Mbps rate, 30 B channels plus one 64-Kbps D channel. Again, it is possible for a customer with lesser requirements to employ fewer B channels, in which case the channel structure is nB + D, where n ranges from 1 to 23, or from 1 to 30 for the two primary services. Also, a customer with high data-rate demands may be provided with more than one primary physical interface. In this case, a single D channel on one of the interfaces may suffice for all signaling needs and the other interfaces may consist solely of B channels (24B or 31B).

The primary interface may also be used to support H channels. Some of these structures include a 64-Kbps D channel for control signaling. When no D channel is present, it is assumed that a D channel on another primary interface at the same subscriber location will provide any required signaling. The following structures are recognized:

- *Primary rate interface H0 channel structures:* This interface supports multiple 384-Kbps H0 channels. The structures are 3H0 + D and 4H0 for the 1.544-Mbps interface, and 5H0 +D for the 2.048-Mbps interface.

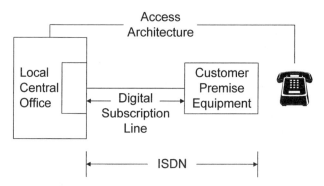

Figure 6.15 The ISDN digital connectivity.

- *Primary rate interface H1 channel structures:* The H11 channel structure consists of one 1,536-Kbps H11 channel. The H12 channel structure consists of one 1,920-Kbps H12 channel and one D channel.

- *Primary rate interface structures for mixtures of B and H0 channels:* Consists of zero or one D channels plus any possible combination of B and H0 channels up to the capacity of the physical interface (for example, 3H0 + 5B + D and 3H0 + 6B).

It should be emphasized that N-ISDN is an *access* standard, not an end-to-end connection standard. In other words, it defines only the connection between the user and the local central office as shown in Figures 6.15 and 6.16.

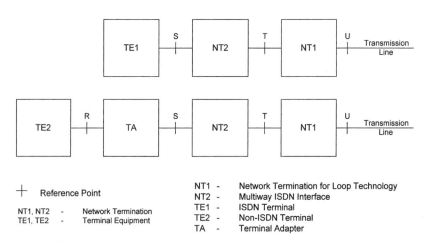

Figure 6.16 ISDN user/network interfaces.

7

ITU-T Recommendation H.323 Packet-Based Multimedia Communications Systems

7.1 Introduction

The ITU-T Recommendation H.323 is a standard that describes systems and equipment for use in providing multimedia communications over packet-based networks. In this context, multimedia communications is generally considered to consist of audio or audiovisual communications for conversational applications. This may also include a data or graphic communications component that is intended to enhance the utility of the audiovisual communications by providing additional tools for collaboration.

While the original development of H.323 was directed toward corporate LANs, the scope has been broadened to any packet-based network such as enterprise area networks, metropolitan area networks, intra-networks, and inter-networks (including the Internet). They also include dial-up connections or point-to-point connections, over general switched telephone network (GSTN) or ISDN, that use an underlying packet-based transport such as point-to-point protocol (PPP). These networks can consist of a single network segment, or they can have complex topologies which incorporate many network segments interconnected by other communications links.

The ITU is the United Nations' Specialized Agency in the field of telecommunications. The ITU-T is a permanent organ of the ITU and is responsible for studying technical, operating, and tariff questions and issuing recommendations on them with a view to standardizing telecommunications on a worldwide basis.

Prior to the development of H.323, the ITU-T had developed other audiovisual communications recommendations including:

- H.320 for operation over N-ISDN networks;
- H.321 for using H.320 equipment on B-ISDN (ATM);
- H.322 for using H.320 equipment on guaranteed quality of service LANs (Iso-Ethernet);
- H.324 for operation of the analog telephone network (GSTN);
- H.310 for operation over B-ISDN (ATM) using MPEG-2.

These recommendations all provide for multimedia communications using a transport mechanism that provides a guaranteed quality of service. In these systems, the transport mechanism provides a well-known and stable data rate, error rate, drop-out rate, delay, and jitter. Many packet-based networks, however, do not have a transport mechanism which guarantees these parameters. In particular, Ethernet-based LANs do not provide a stable or guaranteed data rate. The data rate may fluctuate based on the instantaneous network congestion. In addition, network congestion can cause packet losses, packet arrival time jitter, and delay. These conditions can have a very detrimental effect on audio and video quality. H.323 attempts to take into account these conditions to minimize their effect on the quality of the communications.

There were several goals set out for the development of H.323. These were:

- Provide video teleconferencing capability to the desktop;
- Make use of the existing installed-base of corporate LANs;
- Interoperate with H.320 terminals on the public ISDN network;
- Provide network management with tight control over video teleconferencing traffic;
- Take advantage of network connectivity for multipoint conferences;
- Do not rely on guaranteed quality of service.

Interoperation with the recommendations listed above, and in particular H.320, was of particular importance during the development of H.323. This was accomplished by mandating in H.323 the same audio and video codecs that are mandated in H.320. This provides a minimum mandatory point of interoperability. Optional codecs provide enhanced performance, and gateways can convert from one codec to another, but the mandatory codecs assure that the different terminals will be able to interoperate with minimal delay and complexity in the gateway.

H.323 Version 1 was approved by the ITU-T in November of 1996. Work immediately began to add features and functionality and Version 2 was approved in February 1998. Work continues today to improve and advance the capability of H.323 systems. The following sections describe components defined in H.323, networks aspects, protocols and procedures used, supplementary services, quality of service issues, and security.

7.2 H.323 Components

In addition to the terminal and MCU, H.323 defines several components that had not been used in H.320 systems. These are gatekeepers, gateways, multipoint controllers, and multipoint processors. This section will describe the H.323 components and their use in H.323 systems.

7.2.1 Terminals

The terminal is the basic multimedia communications device specified in H.323. It provides the mechanism for users to communicate verbally, visually, and through data transfer. The mandatory features of the H.323 terminal are similar to those for the H.320 terminal. This was done specifically to provide a minimum level of communications between H.323 and H.320 without having to use complex transcoding gateways. Figure 7.1 shows a functional block diagram of an H.323 terminal.

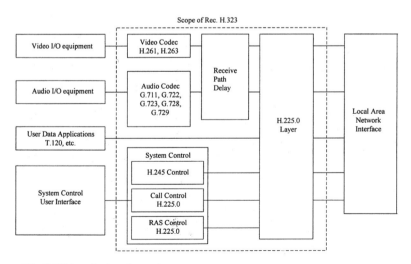

Figure 7.1 H.323 terminal equipment.

G.711 audio coding is mandatory in all H.323 terminals. If it is known that the communications link provides less that 64 Kbps, then either G.723.1 or G.729 is required. Today, most products support G.723.1 audio coding. Video is optional in H.323 systems; however, if it is provided, it must support at least the H.261 QCIF format. Today, most products support the more efficient H.263 video coding. Delay in the receive path is provided for lip sync (synchronizing audio and video) and jitter buffering. Jitter buffering allows for variations in the arrival of audio and video packets from the network. Data communications is also optional. T.120 is the preferred standard for data communications and collaborative applications. This diagram is very similar to the system diagram found in the other H-Series Recommendations except that there is no multiplex layer. Multiplexing of media, control, and data streams is performed by the network by sending different streams to different network "ports." Each port is attached to a process that services the data arriving at that port. In this way, the video process only receives video data.

Several system control functions are provided. Registration Admission Status (RAS) control functions provide communications between the terminal and the gatekeeper for registration, admission, and status functions. This is used to provide tight management of multimedia traffic on the network. Call control provides a mechanism for the packet-based network to act like a switched-circuit network. Functions like dial, ring, answer, and busy are implemented by the call control protocol. The H.245 control function provides for capability exchange, master/slave negotiation, mode request, logical channel control, and other controls and indications.

A key difference between an H.323 terminal and an H.320 terminal is the ability for the H.323 terminal to participate in multipoint conferences without the use of an MCU. Since each terminal on the network can see the traffic from all other terminals on the network, the terminal can perform its own audio mixing and video switching. This is called a distributed multipoint conference as opposed to a centralized multipoint conference which uses an MCU to mix and distribute the audio and video streams.

The H.323 terminal can take many different forms. It can be a powerful video teleconferencing system with audio, video, and data communications. It can receive/transmit a single audio and video stream, or it can handle multiple streams. It can be a simple audio-only Internet telephone. Work is being done today to add real-time fax capability to the H.323 Recommendation. This will allow real-time Internet fax machines.

7.2.2 Gateways

The gateway was defined in order to provide interoperability with H.320 terminals. The gateway provides all of the electrical and logical translation functions that are required to provide communications between an H.323 terminal on the packet-based network and an H.320 terminal on the N-ISDN network. While the original intention was to provide interoperability with H.320 terminals, gateways may provide for interoperability with many other terminal types, including H.324, H.310, and voice-only terminals (telephones) on the GSTN or ISDN. ITU-T Recommendation H.246 specifies the mandatory protocol conversions for many of these gateways. Currently, this recommendation only specifies the H.323 to H.320 protocol conversions; however, work is being done to specify H.323 to voice-only terminal operation.

The gateway operates like an H.323 terminal or MCU on the packet-based network and a switched circuit network (SCN) terminal (i.e., H.320) or MCU on the switched-circuit network, with the protocol conversion functions done in the middle. This is illustrated in Figure 7.2.

The physical implementations of a gateway may vary tremendously. A gateway can be very simple and handle only a single call between the packet-based network and the switched-circuit network. It can also be very complex, handling many simultaneous calls, performing transcoding of the audio and video to assure that every connection participated in its optimal mode, and interfacing to several different switched-circuit networks. It may also include other functions such as the gatekeeper or the MCU.

7.2.3 Gatekeeper

The gatekeeper provides network management functions for H.323 systems. In order to provide strict access controls for corporate networks, all H.323 terminals, MCUs, and gateways must get permission from the gatekeeper before they can make or accept a call. In addition, the gatekeeper gives the terminal permission to use a specific amount of network bandwidth for the call. By having this admission function, complex admission criteria can be developed and policed by the gatekeeper. For example, the gatekeeper may allow a maximum of five audiovisual calls at 128 Kbps each, or it may allow any number of calls up to an aggregate of 2 Mbps. It may prevent calls from some terminals to access gateways (outside calls). Regardless of the strategy, the concept is that the network administrator has a method to control the audiovisual traffic on the packet-based network.

Another function of the gatekeeper is to provide address translation. This provides a directory service which allows users to enter an alias address which the gatekeeper will convert into a network (IP) address for the terminal to use. The alias address may be a telephone number, extension number, name, or other alphanumeric label. This indirect addressing has several benefits. First it provides telephone-like number plans on the packet network. Second, it allows a terminal

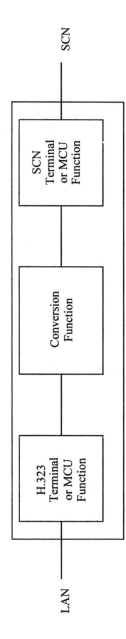

Figure 7.2 Gateway functions.

to move to a new network address without having to let everyone know the new address. The terminal simply registers the new address with the gatekeeper and it will translate the old alias to the new address. This is particularly important for networks which dynamically assign network addresses (i.e., dynamic host control protocol, or DHCP).

Using this address translation mechanism, the gatekeeper can hide system elements from the user to simplify system operation. For example, if a dial prefix of 9 is used to indicate an external call, the gatekeeper can automatically route the call to an available gateway. Without this feature, the user would have to know the address of the gateway and place the call directly to the gateway. This is also useful if a gateway is added or removed. The gatekeeper is updated with the new information, and the users never need to know that a change has been made. This works the same way with MCUs. For each conference, the MCU registers an alias address (conference name) with the gatekeeper. When a user wants to join the conference, it simply calls the conference name, and the gateway routes the call to the proper MCU.

The gatekeeper can also perform other functions such as providing PBX-like functions for a call, keeping accounting information, call logs, and other management functions. An important note is that the gatekeeper is an optional component of an H.323 system. If the gatekeeper is included on the network, the terminals must use it; however, it is not necessary to include the gatekeeper in all systems. If the gatekeeper is not present, the network will not benefit from the tight management controls, and the user will not benefit from the address translation functions.

7.2.4 Multipoint Control Unit

The functions of the traditional MCU have been split into two parts in H.323 systems. The multipoint controller (MC) provides the control functions for the multipoint conference. This includes negotiation of common operating modes, and opening and closing of logical channels for the media stream distribution. The multipoint processor (MP) provide the media stream processing functions such as audio mixing and video switching. When both of these components are present in the same entity, you have an MCU. The primary reason for separating the MC and MP is to allow the MC to be included in the terminal. In this way, terminals can have multipoint conferences without having an MCU. If there are multiple MCs in a conference, a master/slave process is used to determine which one is active for that conference. Each terminal will provide its own audio mixing and video switching functions controlled by the active MC. A by-product of this capability is the ad hoc multipoint conference. This is a conference that starts out as a point-to-point call between two terminals, and then expands into a multipoint conference by inviting another terminal to join.

H.323 defines many types of multipoint conferences. Centralized multipoint conferences use an MCU. Decentralized conferences do not use an MCU. Hybrid multipoint conferences use the centralized model for one media (such as audio mixing) and the decentralized model for the other (video switching). A mixed multipoint conference allows terminals to participate using the centralized model, while other terminals participate using the distributed model. A broadcast conference is one in which there is one speaker and many, possibly thousands, of listeners. A broadcast panel conference is one in which there is a small group in a multipoint conference and many other terminals listening. ITU-T Recommendation H.332 specifies the operation of H.323 terminals in broadcast and broadcast panel conferences.

7.3 Packet-Based Networks

H.323 was originally developed to provide video teleconferencing over the installed base of corporate LANs. These networks provide an existing communications infrastructure which provides connectivity to nearly every corporate desktop. As the standard developed and products were fielded, it was apparent that the scope of H.323 far exceeded the LAN. This is primarily due to the fact that the TCP/IP in network protocols were used in a wide range of network types. These networks range from very simple single segment LANs, to multiple segment LANs with complex topologies, to inter-networks made up of LANs interconnected by wide area links, to the Internet. This section describes the characteristics of either end of that spectrum, the LAN and the Internet.

7.3.1 Local Area Networks

The LAN is a data communications network that interconnects two or more data communicating devices over a transmission medium. A LAN is generally limited in geographic scope to a single building. A LAN is classified by three components: topology, transmission medium, and medium access control techniques.

The most common topologies are star, bus, and ring. The most common mediums are twisted-pair cable, coaxial cable, and fiberoptic cable. 10BaseT Ethernet is an example of a star topology network where each data communications device is connected to a hub over twisted-pair cables. The hub relays the information from one device to all other devices. More complex hubs intelligently relay the information so as to only send information that is addressed to a device to that device. 10Base2 Ethernet is an example of a bus network. In this network, each device is connected by coax cable through a coax "T" connector to a coax backbone. Fiber distributed data interface (FDDI) is an example of ring networks. In these networks, each device receives data over fiberoptic cable from the device before it and transmits data to the device after it. The last device is connected back to the first device. If the data are not addressed to it, the device simply passes them on. Because of the potential for a break in the ring, these networks typically employ redundant rings connected in opposite directions. In this way, if one ring breaks, data can flow around the other ring until it is repaired.

Each of these different types of LANs consists of a collection of devices that share the network's transmission capacity. This provides several advantages as well as drawbacks. One advantage is that data sent in one transmission from one device can be received by several devices. This allows data to be distributed more efficiently, particularly for multipoint communications. Another advantage is that a device can plug into the network at any point and retain its original addressing. This simplifies configuration and flexibility of the network. The drawbacks, however, contribute to the quality of service issues related to H.323. For example, on Ethernet networks, if two devices attempt to transmit at the same time, a collision occurs where the data become corrupted. In some cases this is detected and the data can be retransmitted, which adds delay. In other cases, the data are lost forever, and the receiver must compensate for it in some way. As the traffic on the network increases, these collisions become more frequent, more retransmissions, causing further increasing the congestion. The higher level protocols such as transmission control protocol/Internet protocol (TCP/IP) have mechanisms that attempt to control this runaway congestion.

7.3.2 The Internet

The TCP/IP protocol was originally developed to support the Internet. The Internet is a complex collection of heterogeneous networks interconnected by various backbones and WAN connections. Routers provide the connection between the different network segments, backbones, and access links. The Internet was originally developed to interconnect government agencies and educational institutions to support research activities. Today, the Internet is a worldwide public TCP/IP network providing global-packet data communications not only for government and educational institutions, but also corporations, small businesses, consultants, and the general public.

The Internet is a collection of physical layer networks that support the TCP/IP protocol. The IP protocol provides each endpoint on the Internet with a 32-bit globally unique network address (or IP address). This address consists of a network part that indicates which network the endpoint is on, and a host part that identifies the endpoint on that network. This address is used for routing

packets throughout the Internet between a source and destination. Routing protocols keep the routers within the Internet updated as to the path that packets
should take from source to destination, through the different networks and backbones required to traverse the Internet. The packet structure of TCP/IP (called a
datagram) is shown in Figure 7.3. TCP is the first layer of protocol when sending a message. TCP uses a variable-length packet that can contain up to 65,516
bytes of data. Larger data objects are sent in multiple datagrams. The TCP header
structure is a 20-byte group that includes fields for identifying source and destination for the datagram, a sequence number that allows the receiving TCP to re-
order the datagrams if they arrive out of order, and various other fields and flags
for management of transmission. A checksum field provides error detection for
the contents of the datagram.

An IP header is appended to each datagram generated by TCP. It provides
fields for control of the routing of the datagram through the networks. IP adds
additional source and destination information and also provides fields that support IP processing later in the path to divide the datagram into several pieces if
that will provide best transport. The lifetime field provides for the case where a
datagram might be routed around endlessly and never find its destination. It provides for a maximum number of hops, after which IP will destroy the datagram.

TCP/IP provides an acknowledgment back to the sender for the successful reception of each datagram. This allows lost datagrams to be sent again, or,
after an appropriate timeout, for the sender to know that a message was never
received.

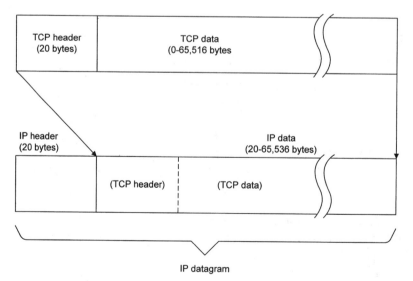

IP datagram

Figure 7.3 TCP/IP packet structure.

Like all other network types which use TCP/IP, the Internet can support H.323 multimedia calls. This means that a person at home can make an H.323 call from a home PC, through the local Internet service provider, through the Internet, to a corporate LAN on the other side of the world, to a computer on that LAN, to a person sitting at that computer terminal. Since Internet traffic is currently free (aside from a small monthly access charge), this has opened a new industry of products and service providers who are using H.323 protocols to provide Internet telephony and fax transmission. By using the Internet, the long distance telephone charges are bypassed.

7.4 Network Services

H.323 was developed to be network independent. The only requirement on the network is that it can provide two classes of data delivery service: guaranteed and best effort. Guaranteed delivery provides a mechanism to assure error-free delivery of the data. Retransmission/acknowledgment techniques are usually used to assure that each packet arrives at the destination error-free. The drawback of this delivery mechanism is that data latency increases because the receiver must buffer the data to allow missing packets to be retransmitted. The transmitter must wait for acknowledgments before sending subsequent packets, packet headers are larger, and initial packets must be exchanged to establish the connection. The benefit is that the data always get there without error. For this reason, this delivery mechanism is used for call establishment messages, conference control messages, and T.120 data where error-free delivery is critical.

Best effort delivery eliminates some of the data latency by eliminating the retransmission/acknowledgment technique. This allows the packet headers to be smaller, and no initial connection establishment is required. The drawback of this delivery mechanism is that packets may be lost due to collisions, errors, and congestion. The receiver must be able to identify that packets were lost, and take action to minimize the impact of the lost data. The benefit of this mechanism is that data are delivered quickly with minimal latency. This delivery mechanism is used for the real-time delivery of audio, video, and remote device control data.

7.4.1 TCP/IP Services

To date, most H.323 systems operate over TCP/IP networks. TCP/IP has become very popular because it is extremely flexible, is independent of the underlying physical network, provides guaranteed delivery and best effort delivery mechanisms, and is the protocol used on the Internet. On a TCP/IP network,

TCP provides the guaranteed delivery mechanism, and user datagram protocol (UDP) provides the best effort delivery mechanism. These protocols are at the transport layer of the ISO network model. They both operate above the IP, which is at the network layer. Below the network layer is the physical/link layer. TCP/IP operates over many different physical/link layers. The IP layer makes the transition between different physical layers transparent to the higher level protocols. This characteristic is allowing TCP/IP protocols to be deployed over almost any physical connection. This includes various shared- and switched-Ethernet LANs, dial-up modem connections, ISDN connections, LAN emulation over ATM networks, satellite links, microwave links, and so on. It is rapidly becoming ubiquitous in packet-based networks of all types. This will make it possible for H.323 to also become ubiquitous on all of these networks, without the need to use gateways.

Above the transport layer are the various tasks that process audio, video, and control data. The transport layer has a mechanism which routes data packets directly to these tasks. In TCP/IP this mechanism is the port number. Each task that is expecting data is assigned a port number. When data are received at the transport level, the port number is examined, and the data are forwarded to the appropriate task for processing. This feature eliminates the need to have a separate multiplex protocol in H.323 systems (such as H.221 in H.320 or H.223 in H.234).

7.5 Protocol Stack

Figure 7.4 illustrates the H.323 protocol stack. The lowest four layers show the network protocols and are outside the scope of H.323. This diagram shows those layers for TCP/IP running on an Ethernet network. The audio and video data streams are packetized according to the real-time transport protocol (RTP) standard. These packets include header information which supports synchronization, loss detection, and status reporting. In addition, a real-time transport control protocol (RTCP) stream is established to provide a channel for passing transmitter and receiver status for the audio and video streams. The RAS channel is used for terminal-to-gatekeeper communications. This provides for registration, admissions, bandwidth control, and status messages. All of these channels are transmitted over the UDP unreliable transport mechanism which provides for low latency, low overhead, and near real-time communications.

The call signaling channel provides the mechanism for establishing a call between two endpoints. This provides telephone-like functions such as call, ring, answer, busy, and hang up on the packet-based network. This channel was modeled after the Q.931 protocol used for call establishment in N-ISDN networks.

Audio Apps	Video Apps	Terminal Control and Management				Data Apps T.126 T.127 T.128 T.13X
G.711 G.722 G.723.1 G.728 G.729	H.261 H.262 H.263	RTCP	H.225.0 RAS Channel	H.225.0 Call Signaling Channel	H.245 Control Channel	T.124 T.125 T.122
RTP				X.224 Class 0		T.123
Unreliable Transport (UDP)				Reliable Transport (TCP)		
Network Layer (IP)						
Link Layer (IEEE 802.3)						
Physical Layer (IEEE 802.3)						

Figure 7.4 H.323 protocol stack.

The H.245 channel provides for end-to-end control of the call or conference. This channel is used for capability negotiations; master/slave determination; mode requests; opening and closing logical channels for audio, video, and data; and other controls and indications. The T.122, T.124, and T.125 provide the multipoint communication service (MCS) and generic conference control (GCC) multipoint transport functions for T.120 data channels. All of these channels are transmitted over the TCP reliable transport mechanism which provides guaranteed error-free delivery of data. The X.224 and T.123 layer defines how these channels are put into TCP packets.

7.6 H.323 Procedures

The first thing that an H.323 terminal needs to do is to determine which gatekeeper to register with. This may be hard coded in the terminal or it may be done through automatic means. These automatic methods are called gatekeeper discovery. A terminal has two ways to find a gatekeeper. If it knows what gatekeepers are on the network, it can query each one using the RAS channel, asking "Are you my gatekeeper?" The terminal can also multicast a query to the network asking "Who is my gatekeeper?" All gatekeepers listen for these messages, and the appropriate gatekeeper will reply.

Once the terminal knows its gatekeeper, it then sends a registration request using the RAS channel, which tells the gatekeeper what its network address is, and any alias addresses or names that it wants to be known by. This is used by the gatekeeper for address translation. Registration can be done once when the terminal is first installed, or it can be done every time the terminal logs into the network. The latter approach is particularly useful in DHCP networks where the terminal may be assigned a different network address every time it logs on. Once the terminal is registered, it is free to make and receive calls.

As shown in Figure 7.5, a terminal wishing to place a call first sends an admission request to the gatekeeper using the RAS channel, asking for permission to place the call. The gatekeeper then determines if the terminal is allowed to place the call based on whatever admission scheme is implemented. The terminal may also send the alias address of the terminal being called. The gatekeeper will translate this alias address into a network address, which then allows the calling terminal to send call establishment messages to the called terminal.

The terminal then sends a setup message, using the call signaling channel, to the called terminal indicating that it wishes to place a call. The called terminal first gets permission from its gatekeeper to accept the call, then returns an alerting message indicating that the phone is ringing. When the user answers the call, the called terminal sends a connect message indicating that the call was answered. New fast-connect procedures allow the audio channel to be connected through immediately so that when the call is answered, the calling and called users can immediately begin to communicate without having to wait for the completion of the remaining protocol exchanges.

Once the call is established, the H.245 control channel is established, and both terminals exchange capability messages indicating what modes they can operate in. Based on these messages, each terminal knows what audio, video, or data streams it may transmit to the other terminal. The terminals then perform a master/slave determination procedure which is used to resolve any conflicts between the terminals. At this point, open logical channel messages can be sent to open

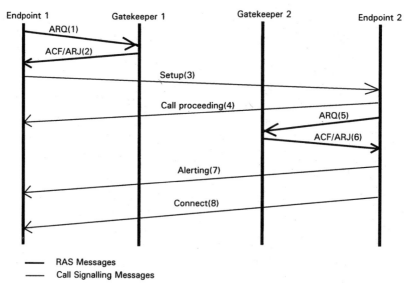

Figure 7.5 Call establishment procedure.

audio, video, or data channels between the two terminals. During the course of the call, audio, video, and/or data channels may be opened and closed, bit rates changed, and operating modes changed using the H.245 control channel.

At the completion of the call, the audio, video, and data logical channels are closed, the call is released, and the gatekeeper is notified that the call has been disengaged. This brings all of the protocol state machines back to the pre-call state, ready for the next outgoing or incoming call.

7.7 Supplementary Services

Many of the procedures and protocols in H.323 are used to make the packet-based network act like the regular telephone network. *Supplementary services* is the name applied to a variety of telephone or PBX services that we use every day on our regular telephone system. The H.450 series of recommendations has been developed to define these services in H.323 systems. This section discusses some of these services.

- *Call transfer:* When terminals A and B are in a call, terminal A may want to transfer terminal B to terminal C. Terminal A will first put terminal B on hold and then call terminal C. Terminal A may first consult with terminal C and then connect terminal B to terminal C or it may immediately connect terminal B to terminal C.

- *Call diversion:* Call diversion is another name for call forwarding. It is the redirection of a call, at the request of the called terminal, prior to answering the call. There are several types of call forwarding. Call Forward—Unconditional routes all calls destined for terminal A to terminal B. Call Forward—Busy routes a call destined for terminal A to terminal B if terminal A is busy. Call Forward—No Reply routes a call destined for terminal A to terminal B if terminal A is does not answer within a certain time. Call Deflection is the same as Call Forward—No Reply except that terminal A continues to ring, and the user can answer at either terminal A or terminal B.

- *Call hold:* When terminal A and terminal B are in a call, terminal A may want to put terminal B on hold with the intention of retrieving terminal B in a short period of time. During this hold condition, terminal B may receive "on-hold" audio and video. Terminal A is free to perform other operations like placing a second call.

- *Call park and pickup:* When terminal A and terminal B are in a call, terminal A may want to park terminal B with the intention of retrieving terminal B from another terminal (C).

- *Call waiting indication:* When terminal A and terminal B are in a call and terminal C attempts to call terminal A, terminal A can receive a call waiting indication. Terminal A then has the option to hang up and answer terminal C, put terminal B on hold and answer terminal C, or ignore the indication.

7.8 Quality of Service Issues

Quality of service (QoS) is generally characterized by data rate and bit error rate. In switched-circuit networks, the quality of service is well known, or guaranteed. For example, in an ISDN BRI connection, the data rate is 128 Kbps, and the bit error rate is very low (better than 1×10^{-8}). The data rate does not change during a call, the clock rate is stable (very little clock jitter), and the bits received at the destination arrive in the same order as they were put in at the source. These characteristics lead to a very clean, predictable communications environment.

In packet-based networks, there is little or no guarantee of the QoS. The data rate may change as the network becomes congested. Packets may be corrupted or lost due to collisions or congestion in the routers. Packets do not arrive at a constant rate, and they may not even arrive in the same order due to routing or retransmission. This unpredictable communications environment could be disastrous for real-time data streams such as audio and video. This might cause the audio frequency to change, cause clicks or hums in audio, cause the video picture to tear or jump, or cause blocks to appear in the video picture.

Because of the potential for poor quality, it was thought that LAN-based multimedia conversational services would be unacceptable to the users. However, nonguaranteed quality of service does not immediately equate to poor QoS. Mechanisms such as buffering can reduce or eliminate the effects of data rate variations, packet arrival time jitter, and sequence errors. Error concealment techniques can be used to minimize or eliminate the effects of packet loss. Within H.323, feedback mechanisms between the sender and receiver can be used to determine the quality of the received media stream. If jitter or packet loss becomes excessive, the sender can choose to reduce its transmission rate to try to compensate for the network congestion. The result is that on well-designed networks, the quality of the audiovisual communications can be excellent. Networks which are congested can be segmented to try to isolate sources of congestion. Network design is critical to the successful deployment of H.323 systems in large networks.

Other work is also taking place to try to improve the QoS on the packet network. The resource reservation protocol (RSVP) is a mechanism to reserve bandwidth through routers within a network. By using this protocol, a multimedia call can reserve a certain amount of the resources of the router. These resources are bandwidth and priority. By setting the priority of the data stream, the packet arrival jitter can be reduced, because the router will service the higher priority packets immediately. This protocol is just beginning to be deployed in routers today. It is important to note that RSVP only controls QoS between network segments. It does nothing to control QoS within a network segment. Fortunately, other work is in progress to develop subnet bandwidth management (SBM) protocols which control the access of each node to the network. This will in effect limit a node's transmission bandwidth and the priority of its access to the network, basically reserving bandwidth and controlling inter-packet jitter within the network segment. This is very new work, and it is hoped that these and other QoS protocols will help to provide guaranteed QoS on the packet-based networks.

7.9 Security and Encryption

Because of the shared media on LANs such as Ethernet, any node on the network may see all of the packets going to or from any other node on the network. This causes concern for the privacy of the communications, and the authenticity of the terminal responding to a call. Placing calls over the Internet was also a cause for concern for the privacy of the call. These issues made H.323 security a high priority, and shortly after the original H.323 version was approved, work began on Recommendation H.235 which specifies mechanisms for providing privacy and authentication in H.323 calls.

Authentication is used to assure that the person that you are connected to is indeed that person. Authentication uses the exchange of digital certificates to accomplish this. These certificates use public key encryption to protect a text string. The receiver of the certificate decrypts it using the sender's public key and if it is the expected text string, then it could only have been sent by the holder of the private key which hopefully is the person the receiver is attempting to call. This authentication method is particularly useful for controlling access to network resources. For example, a gatekeeper can use this as part of the admission criteria to assure that the terminal has authority to use the service that is requested. Similarly, gateways and MCUs can assure that terminals which are connecting to them have authority to do so.

Privacy is used to assure that the communications between two terminals (or other devices) can not be monitored by an unauthorized third party. This is done by encrypting the media streams. First, transport layer security (TLS), which is part of the TCP/IP stack, is used to encrypt the H.245 control channel.

Once this has been done, encryption algorithms can be negotiated and a key exchange can take place between the two terminals. These algorithms are then used to encrypt the data sent over the audio and/or video channels. The encryption algorithms used in these procedures are not specified in H.235 because of the sensitive nature of encryption in the international community. Instead, each community of interest determines which algorithm is best suited for their application. For example, the banking industry may agree on the use of a particular algorithm. They will then buy equipment that can support that algorithm. During capability negotiation, they determine that both (or all) terminals in the conference can support their algorithm, and they will then be able to provide the privacy that they need.

Users must be aware that there may be certain elements in the system where unprotected data reside. For example, a gateway or MCU must decrypt the incoming channels, process the data, and then re-encrypt the outgoing channels. Within the gateway or MCU the data are unencrypted. These devices are called trusted entities and other security measures must be taken to assure the privacy of the data inside. These are usually physical security methods such as locating the equipment in locked, limited-access rooms.

Future work on H.323 security is expected to include real-time integrity and nonrepudiation. Real-time integrity is a mechanism for assuring that what is sent can not be changed without the receiver knowing it. It also involves preventing playback spoofing, where a third party records an encrypted communication and later plays it back to spoof the receiver into releasing critical information such as new keys. Nonrepudiation is similar to authentication. It also uses digital certificates; however, the purpose is to put a signature on a transmission so that the sender of information cannot deny that he or she is the one who sent it. This is used now in electronic commerce to prevent fraud.

8

Multimedia Communications via the PSTN (H.324) and Mobile Radio (H.324M)

8.1 Overview

In September 1993, the ITU established a program to develop an international standard for a videophone terminal operating over the PSTN. A major milestone in this project was accomplished in March 1996, when the ITU approved the standard. It is anticipated that the H.324 [1–5] terminal will have two principal applications: a conventional videophone used primarily by the consumer, and a multimedia system to be integrated into a personal computer for a range of business purposes.

In addition to approving the umbrella H.324 Recommendation, the ITU has completed the four major functional elements of the terminal: the G.723.1 speech coder, the H.263 video coder, the H.245 communication controller, and the H.223 multiplexer. The quality of the speech provided by the new G.723.1 audio coder, when operating at only 6.4 Kbps, is very close to that found on a conventional phone call. The picture quality produced by the new H.263 video coder shows promise of significant improvement relative to many earlier systems. It has been demonstrated that these technical advances, when combined with the high-transmission bit rate of the V.34 modem [4] (33.6 Kbps maximum), yield an overall audiovisual system performance that is significantly improved relative to earlier videophone terminals.

At the same 1996 meeting in Geneva, the ITU announced the acceleration of the schedule to develop a standard for a videophone terminal to operate over mobile radio networks. The new terminal, designated H.324M, will be based upon the design of the H.324 device to ease interoperation between the mobile and telephone networks.

8.2   H.324: Terminal for Low Bit-Rate Multimedia Communication

Recommendation H.324 describes terminals for low bit-rate multimedia communication, utilizing V.34 modems operating over the GSTN. H.324 terminals may carry real-time voice, data, and video, or any combination, including videotelephony.

H.324 terminals can be integrated into personal computers or implemented in stand-alone devices such as videotelephones. Support for each media type (such as voice, data, and video) is optional, but if supported, the ability to use a specified common mode of operation is required so that all terminals supporting that media type can interwork. H.324 allows more than one channel of each type to be in use. Other recommendations in the H.324 series include the H.223 multiplex, H.245 control, H.263 video codec, and G.723.1 audio codec.

H.324 makes use of the logical channel signaling procedures of Recommendation H.245, in which the content of each logical channel is described when the channel is opened. Procedures are provided for expression of receiver and transmitter capabilities, so that transmissions are limited to what receivers can decode; receivers can request a particular desired mode from transmitters. Since the procedures of H.245 are also planned for use by Recommendation H.310 for ATM networks and Recommendation H.323 for packetized networks, interworking with these systems should be straightforward.

H.324 terminals can be used in multipoint configurations through MCUs and can interwork with H.320 terminals on the ISDN as well as with terminals on wireless networks. H.324 implementations are not required to have each functional element, except for the V.34 modem, H.223 multiplex, and H.245 system control protocol, which shall be supported by all H.324 terminals.

H.324 terminals offering audio communication will support the G.723.1 audio codec. H.324 terminals offering video communication will support the H.263 and H.261 video codecs. H.324 terminals offering real-time audiographic conferencing should support the T.120 protocol suite. In addition, other video and audio codecs and other data protocols can optionally be used via negotiation over the H.245 control channel.

If a modem external to the H.324 terminal is used, terminal/modem control shall be according to V.25ter.

Multimedia information streams are classified into video, audio, data, and control as follows.

- Video streams are continuous traffic-carrying moving color pictures. When used, the bit rate available for video streams may vary according to the needs of the audio and data channels.

- Audio streams are real time but may optionally be delayed in the receiver processing path to maintain synchronization with the video streams. To reduce the average bit rate of audio streams, voice activation may be provided.

- Data streams may represent still pictures, facsimile, documents, computer files, computer application data, undefined user data, and other data streams.

- Control streams pass control commands and indications between remote-like functional elements. Terminal-to-modem control is according to V.25ter for terminals using external modems connected by a separate physical interface. Terminal-to-terminal control is according to H.245.

The H.324 document refers to other ITU recommendations, as illustrated in Figure 8.1, that collectively define the complete terminal. Four new companion recommendations include H.263 (Video Coding for Low Bit-Rate Communication), G.723.1 (Speech Coder for Multimedia Telecommunications Transmitting at 5.3 and 6.3 Kbps), H.223 (Multiplexing Protocol for Low Bit-Rate Multimedia Terminals), and H.245 (Control of Communications Between Multimedia Terminals). H.324 specifies use of the V.34 modem, which operates up to 33.6 Kbps, and the V.8 (or V.8bis) procedure to start and stop data transmission. An optional data channel is defined to provide for the exchange of computer data in the workstation/PC environment. H.324 specifies the use of the T.120 protocol as one possible means for this data exchange. Recommendation H.324 defines the seven phases of a call: setup, speech only, modem training, initialization, message, end, and clearing.

To be compliant with the standard, V. 34 modems must be used to communicate the H.324 bitstreams between terminals. To minimize market entry cost, most of the initial H.324 implementations employ a PC platform as shown in Figure 8.2. Like all signals obtained from a standard PC serial port, the H.324 data is available asynchronously to be fed to the V.34 modem. However, the output of the V.34 modem is a synchronous bitstream. V.80 is a standard protocol which is used to adapt the V.34 synchronous bitstream for transfer over the asynchronous RS232 port.

8.3   G.723.1: Speech Coder for Multimedia Telecommunications Transmitting at 5.3 and 6.3 Kbps

All H.324 terminals offering audio communication will support both the high and low rates of the G.723.1 audio codec. G.723.1 receivers will be capable of

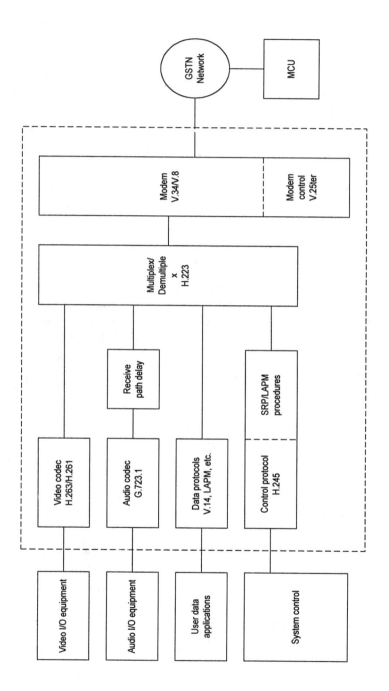

Figure 8.1 Block diagram for the H.324 multimedia system.

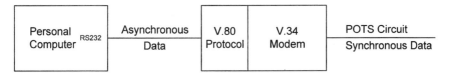

Figure 8.2 Use of V.80 in H.324 multimedia terminals.

accepting silence frames. The choice of which rate to use is made by the transmitter and is signaled to the receiver in-band in the audio channel as part of the syntax of each audio frame. Transmitters may switch G.723.1 rates on a frame-by-frame basis, or based on bit rate, audio quality, or other preferences. Receivers may signal, via H.245, a preference for a particular audio rate or mode.

Alternative audio codecs may also be used, via H.245 negotiation. Coders may omit sending audio signals during silent periods after sending a single frame of silence or may send silent background fill frames if such techniques are specified by the audio codec recommendation in use.

More than one audio channel may be transmitted, as negotiated via the H.245 control channel.

The G.723.1 speech coder can be used for a wide range of audio signals but is optimized to code speech. The system's two mandatory bit rates are 5.3 and 6.3 Kbps. The coder is based on the general structure of the MP-MLQ speech coder. The MP-MLQ excitation will be used for the high-rate version of the coder. The algebraic codebook excitation linear prediction (ACELP) excitation is used for the low-rate version. The coder provides a quality essentially equivalent to that of a POTS toll call. For clear speech, or with background speech, the 6.3-Kbps mode provides speech quality equivalent to the 32-Kbps G.726 coder. The 5.3-Kbps mode performs better than the IS54 digital cellular standard.

Performance of the coder has been demonstrated by extensive subjective testing. The speech quality in reference to 32-Kbps G.726 ADPCM (considered equivalent to toll quality) and 8-Kbps IS54 VSELP is given in Table 8.1. This table is based on a subjective test conducted for the French language. In all cases, the performance of G.723.1 was rated better than, or equal to, IS54. All tests were conducted with four talkers except the speaker variability test, for which twelve talkers were used. The symbols $<$, $=$, and $>$ are used to identify less than, equivalent to, and better than, respectively. Comparisons were made by taking into account the statistical error of the test. The background noise conditions are speech signals mixed with the specified background noise.

From these results one can conclude that within the scope of the test, both low- and high-rate coders are always equivalent to, or better than, IS54,

Table 8.1
Results of Subjective Test for G.723.1

	High Rate	Low Rate
Speaker variability	= G.726	= IS54
One encoding	= G.726	= IS54
Tandem	> 4*G.726	> 4*G.726
Level +10 dB	= G.726	= G.726
Level −10 dB	> G.726	> G.726
Frame erasures (3%)	= G.726-0.5	= G.726-0.75
Flat input (loudspeaker)	= G.726	= G.726
Flat in (loudspeaker) 2T	> 4*G.726	> 4*G.726
Office noise (18 dB) DMOS	= IS54	= IS54
Babble noise (20 dB) DMOS	= G.726	= IS54
Music noise (20 dB) DMOS	> IS54	= IS54

except for the low-rate coder with music. The high-rate coder is always equivalent to G.726, except for office and music background noises.

The complexity of the dual rate coder depends on the DSP chip and the implementation, but it is approximately 18 and 16 MIPS for 6.3 and 5.3 Kbps, respectively. The memory requirements for the dual rate coder are

RAM: 2,240 16-bit words;

ROM: 9,100 16-bit words (tables);

7,000 16-bit words (program).

The algorithmic delay is 30-ms frame plus 7.5-ms look ahead, resulting in 37.5 ms.

G.723.1 can be integrated with any voice activity detector to be used for speech interpolation or discontinuous transmission schemes. Any possible extensions would need agreements on the proper procedures for encoding low-level background noises and comfort noise generation.

8.4 H.263: Video Coding for Low Bit-Rate Communication

All H.324 terminals offering video communication will support both the H.263 [5, 6] and H.261 video codecs, except H.320 interworking adapters, which are not terminals and do not have to support H.263 (see Section 8.2). The H.261 and H.263 codecs will be used without BCH error correction and without error-correction framing. The five standardized image formats are 16CIF, 4CIF, CIF, QCIF, and SQCIF (see Table 8.2).

CIF and QCIF are defined in H.261. SQCIF, 4CIF, and 16CIF are defined in H.263. For the H.261 algorithm, SQCIF is any active picture size less than QCIF, filled out by a black border, and coded in the QCIF format. For all these formats, the pixel aspect ratio is the same as that of the CIF format. Table 8.2 shows which picture formats are required and which are optional for H.324 terminals that support video.

All video decoders shall be capable of processing video bitstreams of the maximum bit rate that can be received by the implementation of the H.223 multiplex (for example, maximum V.34 rate for single link and 2 × V.34 rate for double link).

Which picture formats, minimum number of skipped pictures, and algorithm options can be accepted by the decoder are determined during the capability exchange using H.245. After that, the encoder is free to transmit anything that is in line with the decoder's capability. Decoders that indicate capability for a particular algorithm option will also be capable of accepting video bitstreams that do not make use of that option.

The H.263 coding algorithm is an extension of H.261. H.263 describes, as H.261 does, a hybrid DPCM/DCT video coding method. Both standards

Table 8.2
Picture Formats for Video Terminals

Picture Format	Luminance Pixels	Encoder		Decoder	
		H.261	H.263	H.261	H.263
SQCIF	128 × 96 for H.263‡	Optional‡	Required*†	Optional‡	Required*
QCIF	176 × 144	Required	Required*†	Required	Required*
CIF	352 × 288	Optional	Optional	Optional	Optional
4CIF	704 × 576	Not defined	Optional	Not defined	Optional
16CIF	1,408 × 1,152	Not defined	Optional	Not defined	Optional

* Optional for H.320 interworking adapters

† It is mandatory to encode one of the picture formats QCIF and SQCIF; it is optional to encode both formats.

‡ H.261 SQCIF is any active size less than QCIF, filled out by a black border and coded in QCIF format.

use techniques such as DCT, motion compensation, variable length coding, and scalar quantization and both use the well-known macroblock structure. Differences between H.263 and H.261 are:

- H.263 has an optional GOB level;
- H.263 uses different VLC tables at the macroblock and block levels;
- H.263 uses half-pel motion compensation instead of full-pel plus loop filter;
- In H.263, there is no still picture mode (JPEG is used for still pictures);
- In H.263, there is no error detection/correction included like the BCH in H.261;
- H.263 uses a different form of macroblock addressing;
- H.263 does not use the end of block marker.

It has been shown that the H.263 system typically outperforms H.261 (when adapted for the GSTN application) by 2.5 to 1. This means that when adjusted to provide equal picture quality, the H.261 bit rate is approximately 2.5 times that for the H.263 codec.

The basic H.263 standard also contains the five important optional annexes which are listed below. Annexes D through G are particularly valuable for the improvement of picture quality.

Annex D	Unrestricted Motion Vector;
Annex E	Syntax-Based Arithmetic Coding;
Annex F	Advanced Prediction;
Annex G	PB-Frames;
Annex H	Forward Error Correction for Coded Video Signal.

Of particular interest is the optional PB-frame mode. A PB-frame consists of two pictures being coded as one unit. The name *PB* comes from the name of picture types in MPEG, where there are P-pictures and B-pictures. Thus, a PB-frame consists of one P-picture that is predicted from the last decoded P-picture, and one B-picture that is predicted both from the last decoded P-picture, and the P-picture currently being decoded. This last picture is called a B-picture because parts of it may be bidirectionally predicted from the past and future P-pictures. The prediction process is illustrated in Figure 8.3.

In January 1998, the ITU approved the following additional 12 optional annexes to H.263. These annexes are informally known as H.263+ and constitute Version 2 of H.263.

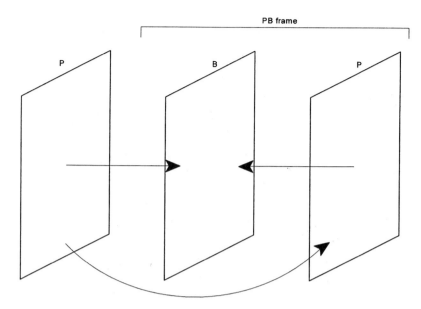

Figure 8.3 Prediction in PB-frames mode.

Annex I	Advanced Intra Coding;
Annex J	Deblocking Filter;
Annex K	Slice Structure Mode;
Annex L	Supplemental Enhancement Information Specification;
Annex M	Improved PB-Frames;
Annex N	Reference Picture Selection;
Annex O	Temporal, SNR, and Spatial Scalability;
Annex P	Reference Picture Resampling;
Annex Q	Reduced-Resolution Update;
Annex R	Independently Segmented Decoding;
Annex S	Alternative Inter VLC;
Annex T	Modified Quantization.

As indicated above, Version 2 of H.263 has a total of 17 optional coding modes as defined by annexes D through T. Although this represents a rich array of advanced video coding technology which will result in a significant improvement in features, picture quality, and error robustness, it also complicates the interoperability issue. To improve this potential interoperability problem, the ITU has appended Appendix III to H.263 (Recommended Optional Enhancement) which defines three levels of preferred groupings of annexes. The preferred levels are summarized in Table 8.3.

Level 1 consists of a set of four annexes which are relatively simple to implement and yield significant improvement in video coding performance. Levels 2 and 3 consist of annexes which are progressively more complex to implement and/or yield less performance improvement.

8.5   H.245: Control Protocol for Multimedia Communications

The control channel carries end-to-end control messages governing the operation of the H.324 system, including capabilities exchange, opening and closing of logical channels, mode preference requests, multiplex table entry transmission, flow control messages, and general commands and indications.

There is exactly one control channel in each direction within H.324, which uses the messages and procedures of Recommendation H.245 [7]. The control channel is carried on logical channel 0 and is considered to be permanently open from the establishment of digital communication until the termination of digital communication. The normal procedures for opening and closing logical channels does not apply to the control channel.

General commands and indications are chosen from the message set contained in H.245. In addition, other command and indication signals can be sent that have been specifically defined to be transferred in-band within video, audio, or data streams (see the appropriate recommendation to determine if such signals have been defined).

Table 8.3
H.263 Preferred Modes

Annex		
Level 1	I	Advanced Intra Coding
	J	Deblocking Filter
	L (L.4)	Full Frame Freeze
	T	Modified Quantization
Level 2	D	Unrestricted Motion Vector
	K	Slice Structure
	P	Reference Picture Resampling
Level 3	F	Advanced Prediction
	M	Improved PB-Frames
	R	Independent Segment Decoding
	S	Alternate Inter VLC

H.245 messages fall into four categories: request, response, command, and indication. Request messages require a specific action by the receiver, including an immediate response. Response messages respond to a corresponding request. Command messages require a specific action but do not require a response. Indication messages are informative only and do not require any action or response. H.324 terminals will respond to all H.245 commands and requests as specified in H.245 and will transmit accurate indications reflecting the state of the terminal.

Table 8.4 shows how the total bit rate available from the modem might be divided into its various constituent virtual channels by the H.245 control system. The overall bit rates are those specified in the V.34 modem. Note that V.34 can operate at increments of 2.4 Kbps, up to 36.6 Kbps. Speech is shown for two bit rates that are representative of possible speech coding rates. The video bit rate shown is what is left after deducting the speech bit rates from the overall transmission bit rate. The data take a variable number of bits from the video, either a small amount or all the video bits, depending on the designer's or the user's control. Provision is made for both point-to-point and multipoint operation. Recommendation H.245 creates a flexible, extensible infrastructure

Table 8.4
Example of a Bit-Rate Budget for Very Low Bit-Rate Visual Telephony
(Kilobits per Second)

	Modem Bit Rate	Virtual Channel		
		Speech	Video	Data
	9.6 Kbps	5.3 Kbps	4.3 Kbps	Variable
Overall transmission bit rate	14.4	5.3	9.1	Variable
	† :	:	:	
	28.8	6.3	22.5	Variable
	33.6	6.3	27.3	Variable
Virtual channel bit rate characteristic		Dedicated, fixed bit rate*	Variable bit rate	Variable bit rate
Priority‡		Highest priority	Lowest priority	Higher than video, lower than overhead/speech

* The plan includes consideration of advanced speech codec technology such as a dual bit-rate speech codec and a reduced bit rate when voiced speech is not present.

† V.34 operates at increments of 2.4 Kbps, that is, 16.8, 19.2, 21.6, 24.0, 26.4, 28.8, 33.6 Kbps.

: The channel priorities will not be standardized; the priorities indicated are examples.

for a wide range of multimedia applications including storage/retrieval, messaging, and distribution services as well as the fundamental conversational use. The control structure is applicable to the situation where only data and speech are transmitted (without motion video) as well as the case where speech, video, and data are required.

8.6   H.223: Multiplexing Protocol for Low Bit-Rate Multimedia Communication

This recommendation [8] specifies a packet-oriented multiplexing protocol designed for the exchange of one or more information streams between higher layer entities such as data and control protocols and audio and video codecs that use this recommendation.

In this recommendation, each information stream is represented by a unidirectional logical channel that is identified by a unique logical channel number (LCN). LCN 0 is a permanent logical channel assigned to the H.245 control channel. All other logical channels are dynamically opened and closed by the transmitter using the H.245 OpenLogicalChannel and CloseLogicalChannel messages. All necessary attributes of the logical channel are specified in the OpenLogicalChannel message. For applications that require a reverse channel, a procedure for opening bidirectional logical channels is also defined in H.245. The general structure of the multiplexer is shown in Figure 8.4. The multiplexer consists of two distinct layers, particularly, a multiplex (MUX) layer and an adaptation layer (AL).

8.6.1   Multiplex Layer

The MUX layer is responsible for transferring information received from the AL to the far end using the services of an underlying physical layer. The MUX layer exchanges information with the AL in logical units called MUX-SDUs, which always contain an integral number of octets that belong to a single logical channel. MUX-SDUs typically represent information blocks whose start and end mark the location of fields that need to be interpreted in the receiver.

MUX-SDUs are transferred by the MUX layer to the far end in one or more variable length packets called MUX-PDUs. MUX-PDUs consist of the HDLC opening flag, followed by a one-octet header and by a variable number of octets in the information field that continue until the closing HDLC flag (see Figures 8.5 and 8.6). The HDLC zero-bit insertion method is used to ensure that a flag is not simulated within the MUX-PDU.

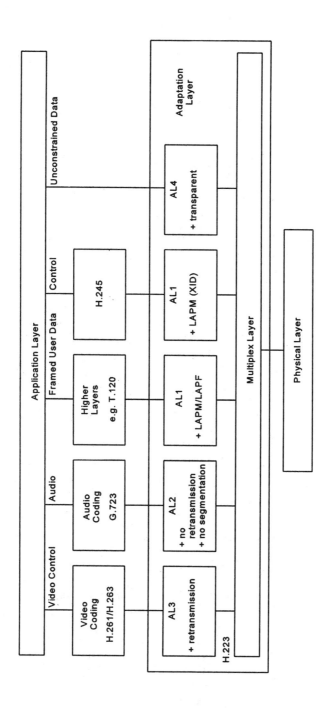

Figure 8.4 Protocol structure of H.223.

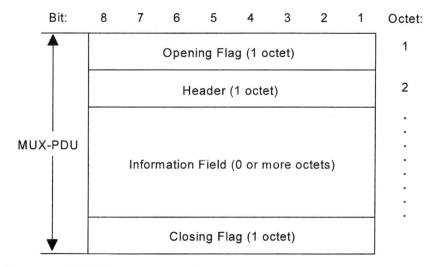

Figure 8.5 MUX-PDU format.

Octets from multiple logical channels may be present in a single MUX-PDU information field. The header octet contains a 4-bit multiplexes code (MC) field that specifies, by reference to a multiplex table entry, the logical channel to which each octet in the information field belongs. Multiplex table entry 0 is permanently assigned to the control channel. Other multiplex table entries are formed by the transmitter and are signaled to the far end via the control channel prior to their use.

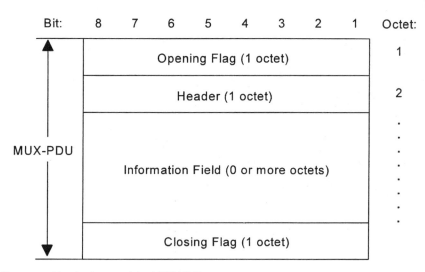

Figure 8.6 Header format of the MUX-PDU.

Multiplex table entries specify a pattern of slots each assigned to a single logical channel. Any one of 16 multiplex table entries may be used in any given MUX-PDU. This allows rapid low-overhead switching of the number of bits allocated to each logical channel from one MUX-PDU to the next. The construction of multiplex table entries and their use in MUX-PDUs is entirely under the control of the transmitter, subject to certain receiver capabilities.

8.6.2 Adaptation Layer

The unit of information exchanged between the AL and the higher-layer AL users is an AL-SDU. The method of mapping information streams from higher layers into AL-SDUs is outside the scope of this recommendation and is specified in the system recommendation that uses H.223.

AL-SDUs contain an integer number of octets. The AL adapts AL-SDUs to the MUX layer by adding, where appropriate, additional octets for purposes such as error detection, sequence numbering, and retransmission. The logical information unit exchanged between peer AL entities is called an AL-PDU. An AL-PDU carries exactly the same information as a MUX-SDU.

Three different types of ALs, named AL1 through AL3, are specified in this recommendation.

AL1 is designed primarily for the transfer of data or control information. Since AL1 does not provide any error control, all necessary error protection should be provided by the AL1 user. In the framed transfer mode, AL1 receives variable length frames from its higher layer (for example, a data link layer protocol such as LAPM/V.42 or LAPF/Q.922, which provides error control) in AL-SDUs and simply passes these to the MUX layer in MUX-SDUs without any modifications. In the unframed mode, AL1 is used to transfer an unframed sequence of octets from an AL1 user. In this mode, one AL-SDU represents the entire sequence and is assumed to continue indefinitely.

AL2 is designed primarily for the transfer of digital audio. AL2 receives frames, possibly of variable length, from its higher layer (for example, an audio encoder) in AL-SDUs and passes these to the MUX layer in MUX-SDUs, after adding one octet for an 8-bit CRC and optionally adding one octet for sequence numbering.

AL3 is designed primarily for the transfer of digital video. AL3 receives variable length frames from its higher layer (for example, a video encoder) in AL-SDUs and passes these to the MUX layer in MUX-SDUs, after adding two octets for a 16-bit CRC and optionally adding one or two control octets. AL3 includes a retransmission protocol designed for video.

An example of how audio, video, and data fields could be multiplexed by the H.223 systems is illustrated in Figure 8.7.

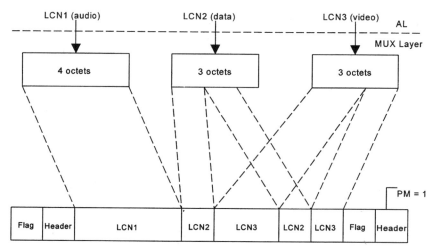

Figure 8.7 Information field example.

8.7 Data Channel

All data channels are optional. Standardized options for data applications include:

- T.120 series for point-to-point and multipoint audiographic teleconferencing including database access, still image transfer and annotation, application sharing, and real time file transfer;
- T.84 (SPIFF) point-to-point still image transfer cutting across application borders;
- T.434 point-to-point telematic file transfer cutting across application borders;
- H.224 for real-time control of simplex applications, including H.281 far end camera control;
- Network link layer, per ISO/IEC TR9577 (supports IP and PPP network layers, among others);
- Unspecified user data from external data ports.

These data applications may reside in an external computer or other dedicated device attached to the H.324 terminal through a V.24 or equivalent interface (implementation dependent), or may be integrated into the H.324 terminal itself. Each data application makes use of an underlying data protocol for link layer transport. For each data application supported by the H.324 terminal, this recommendation requires support for a particular underlying data protocol to ensure interworking of data applications.

The H.245 control channel is not considered a data channel. Standardized link layer data protocols used by data applications include:

- Buffered V.14 mode for transfer of asynchronous characters, without error control;
- LAPM/V.42 for error-corrected transfer of asynchronous characters (additionally, depending on application, V.42bis data compression may be used);
- HDLC frame tunneling for transfer of HDLC frames;
- Transparent data mode for direct access by unframed or self-framed protocols.

All H.324 terminals offering real-time audiographic conferencing should support the T.120 protocol suite.

8.8 Extension of H.324 to Mobile Radio (H.324M)

In February 1995, the ITU requested that the LBC Experts Group begin work to adapt the H.324 series of GSTN recommendations for application to mobile networks. It is generally agreed that a very large market will develop in the near future for mobile multimedia systems. Laptop computers and hand-held devices are already being configured for cellular connections. The purpose of the H.324M standard is to enable the efficient communication of voice, data, still images, and video over such mobile networks. It is anticipated that there will be some use of such a system for interactive videophone applications for use when people are traveling. However, it is expected that the primary application will be nonconversational where the mobile terminal would be usually receiving information from a fixed remote site. Typical recipients would be a construction site, police car, automobile, and repair site. On the other hand, it is expected that there will be a demand to send images and video *from* a mobile site such as insurance adjuster, surveillance, repair site, train, construction site, and fire scene. The advantage of noninteractive communication of this type is that the transmission delay can be relatively large without being noticed by the user.

Several of the general principles and underlying assumptions upon which the H.324M Recommendations have been based are as follows:

- H.324M Recommendations should be based upon H.324 as much as possible.

- The technical requirements and objectives for H.324M are essentially the same as for H.324.
- Since the vast majority of mobile terminal calls are with terminals in fixed networks, it is very important that H.324M recommendations be developed to maximize interoperability with these fixed terminals.
- It is assumed that the H.324M terminal has access to a transparent/synchronous bitstream from the mobile network.
- It is proposed to provide the manufacturer of mobile multimedia terminals with a number of optional error protection tools to address a wide range of mobile networks, that is, regional and global, present and future, and cordless and cellular. Consequently H.324M tools should be flexible, bit rate scalable, and extensible to the maximum degree possible.
- Like H.324, nonconversational services are an important application for H.324M.

Work toward the H.324M recommendation has been divided into the following areas of study: (1) speech error protection, (2) video error protection, (3) communications control (adjustments to H.245), (4) multiplex/error control of the multiplexed signal, and (5) system.

Table 8.5 is a summary of the standardization work which has been accomplished by the ITU to extend the H.324 POTS Recommendation to specify H.324M for the mobile environment. The H.324M standard in each of the technical areas listed above is summarized below.

1. *Audio:* In 1996, the ITU added Annex C (Scalable Channel Coding Scheme for Wireless Applications To Be Used With Multimedia Communication Transmitting at 5.3 and 6.3 Kbps) to the G.723.1 standard to improve the robustness of the basic G.723.1 speech signal to better deal with transmission errors. To develop the standard the error sensitivity of the various elements of the G.723.1 signal was determined, and the error protection of Annex C is applied unequally with the greatest protection associated with the most sensitive bits. In addition, the Annex C error protection is automatically bit rate scalable so that the transmission bit rate, whatever it may be, is used most effectively. That is, as the transmitted bit rate is increased, the error protection is automatically increased proportionately.

2. *Video:* The error robustness of the basic H.263 video signal [9–11] has been improved for the mobile application by the addition of Appendix II (Error Tracking), Annex N (Reference Picture Selection (RPS)),

Table 8.5
Extension of H.324 to Mobile (H.324M)

	H.324 (POTS)	H.324M (Mobile)	Approved
System	H.324	H.324 Annex C	1/98
Audio	G.723.1	G.723.1 Annex C - Bit Rate Scalable Error Protection - Unequal Error Protection	5/96
Video	H.263	H.263 Appendix II—Error Tracking Annex K—Slice Structure Annex N—Reference Picture Selection Annex R—Independent Segmented Decoding	
Communication Control	H.245	Mobile Code Points	1/98
Multiplex	H.223	H.223 Annexes A—increase sync flag from 8 to 16 bits B—A + more robust header C—B + more robust payload	1/98

Annex R (Independent Segmented Decoding), and Annex K (Slice Structure Mode). Appendix II describes the situation in which the decoder determines that a received GOB is in error. Unless corrected, this GOB will be used for the prediction of GOBs in future frames, and the effect of the error will multiply. To avoid this error magnification, a message is sent to the encoder requesting that the GOB in error be transmitted in the intra mode.

3. The basic H.263 signal defines a rigid GOB structure consisting of an array of 11 × 3 macroblocks—three lines of 11 macroblocks per line. A synchronization code separates GOBs so that the effect of a transmission error can be restricted to a GOB rather than impacting the entire picture. However, a rigid GOB structure is not ideal from an error sensitivity perspective because the number of transmitted bits per GOB varies over a very wide range. To improve H.263 error robustness, Annex K provides for a flexible GOB, or slice, structure so that the number of transmitted bits per GOB/slice can be equalized and thereby be given equal error protection.

4. When basic motion compensation is employed in H.263, there is an interactive dependence between adjacent GOBs, particularly in the vertical direction. This occurs because motion prediction can cross

GOB boundaries. Annex R (Independent Segmented Decoding) eliminates this dependence of adjacent GOBs by coding each GOB as if it were an independent picture, thereby reducing the propagation of transmission errors from one GOB to another.

5. Annex N is similar to the error-tracking approach in that it relies on a feedback channel to efficiently stop error propagation after transmission errors. However, instead of transmitting INTRA-coded blocks to clean up the errors, the RPS mode allows the encoder to select one of several previously decoded frames as a reference picture for prediction. RPS can be operated in two different modes. In the ACK mode, all correctly received GOBs are acknowledged, and the encoder only uses acknowledged GOBs as a reference. In the NACK mode, only erroneously received GOBs are signaled by sending NACKs.

6. *Multiplex:* As explained in Section 8.6, the H.223 multiplexer generates a packetized bitstream for transmission over the POTS network. This bitstream is well suited for the POTS network because the telephone system is not extremely error prone. However, the basic H.223 bitstream is not well designed for mobile networks where the bit error rate can be extremely high. To adapt the H.223 bitstream for the mobile environment, the ITU has recently approved optional Annexes A, B, and C to provide three different levels of error control.

7. Figure 8.4 illustrates the structure of the basic H.223 transmitted packet. Note that a single-octet field is assigned to the opening and closing flag. Since it is extremely important to receive these flags correctly, Annex A defines a very simple modification to H.223 to double the length of the flags to make them more robust to errors. The more robust sync flag is 0100 1101 1110 0001, and the payload is not bit stuffed. Similarly, Annex B increases the length of the 1-octet header field to reduce its error sensitivity in addition to the flags. The improved header also includes a multiplex payload length (MPL) field which defines the length of the payload in bytes. This provides additional redundancy by informing the receiver where the next sync flag can be expected. Finally, Annex C adds forward error control features to the adaptation layer of the H.223 multiplexer described in Section 8.6.2 so that the robustness of the payload itself is improved.

8. *Communication control:* As described in Section 8.5, H.245 is the flexible standard defining end-to-end control messages for the H.324 system. It is clear that many new messages are required to implement the additional functions which have been added by the H.324M mobile standard. For example, messages have been added to signal for the

activation, or deactivation, of any of the H.324M optional modes of operation.

9. *System:* To implement H.324M at the system level, H.324 Annex C (Multimedia Telephone Terminals Over Error Prone Channels) has been added. The primary purpose of the annex is to refer to all the other elements of H.324M. However, it also defines the procedure to dynamically change from one level of error control to another during a session.

8.9 Extension of H.324 to N-ISDN

The H.320 standard, providing for multimedia communications over N-ISDN, was approved by the ITU in 1990 and has served as the cornerstone for video-conferencing since that date. However, the ITU has recognized that the technology represented in the new H.324 standard, approved in 1996, is significantly superior to H.320 and directly applicable to the N-ISDN. Therefore, in January 1998, the ITU added Annex D (Operation on ISDN Circuits—H.324/I) to the H.324 standard defining the operation of H.324 for operation on ISDN circuits. The annex defines calling procedures to insure smooth interconnection between H.324 and H.320 terminals on POTS and ISDN networks. It is anticipated that H.324 will slowly replace H.320 systems for ISDN operation in the future.

References

[1] ITU-T Rec. H.324, "Terminal for Low Bitrate Multimedia Communication," 1995.

[2] Lindberg, D., and H. Malvar, "Multimedia Teleconferencing with H.324," *Standards and Common Interfaces for Video Information Systems*, K.R. Rao, ed., Bellingham, WA: SPIE Optical Engineering Press, 1995, pp. 206–32.

[3] Lindberg, D., "The H.324 Multimedia Communication Standard," *IEEE Commun. Mag.*, Vol. 34, no. 12, pp. 46–51, Dec. 1996.

[4] ITU-T Rec. V.34, "A Modem Operating at Data Signaling Rates of Up to 28,8000 bit/s for Use on the General Switched Telephone Network and on Leased Point-to-Point 2-Wire Telephone-Type Circuits," 1994.

[5] ITU-T Rec. H.263, "Video Coding for Low Bitrate Communication," 1996.

[6] Girod, B., N. Färber, and E. Steinbach, "Performance of the H.263 Video Compression Standard," *J. VLSI Signal Processing: Sys. For Signal Image, and Video Tech.*, Vol. 17, Nov. 1997, pp. 101–111.

[7] ITU-T Rec. H.245, "Control Protocol for Multimedia Communication," 1996.

[8] ITU-T Rec. H.223, "Multiplexing Protocol for Low Bitrate Multimedia Communication," 1996.

[9] Park, J. W., J.W. Kim, and S.U. Lee, "DCT Coefficient Recovery-Based Error Conceal-ment Technique and its Application to MPEG-2 Bit Stream Error," *IEEE Trans. Circuits and Sys. for Video Technology*, Vol. 7, no. 6, Dec. 1997, pp. 845–854.

[10] Wen, G., and J. Villasenor, "A Class of Reversible Variable Length Codes for Robust Image and Video Eoding," *IEEE Int'l. Conf. Image Proc.*, Vol. 2, 1997, pp. 65–68.

[11] Steinbach, E., N. Färber, and B. Girod, "Standard Compatible Extension of H.263 for Robust Video Transmission in Mobile Environments," *IEEE Trans. Circuits and Sys. for Video Tech.*, Vol. 7, no. 6, Dec. 1997, pp. 872–881.

9

Multimedia Transmission via B-ISDN (H.321 and H.310)

9.1 Broadband ISDN

The ISDN communication network standard is divided into narrowband and broadband parts. Narrowband ISDN, or N-ISDN, operates at rates equal to or less than the primary rates (for example, 1.544 Mbps), while broadband ISDN, or B-ISDN, operates at rates above the primary rates.

Broadband aspects of the ISDN were developed by the ITU-T to establish a customer-switched digital network. The network node interface (NNI) was standardized by a worldwide unique synchronous digital hierarchy (SDH). Figure 9.1 illustrates the worldwide unique NNI. The SDH specifies 155.52 Mbps as the worldwide unique interface bit rate. As described in Recommendation I.121, the target transfer mode is the asynchronous transfer mode (ATM) in which the data is transmitted in a series of fixed-size blocks called cells (see Figure 9.2). Packet-switched networks already exist for the transmission of digital data for nonreal-time services (for example, the exchange of information between computer databases). In this instance, if a packet is corrupted or lost, the receiving terminal can request that the particular packet be retransmitted. Recommendation I.121, however, envisages that the B-ISDN will carry all the telecommunication services such as telephony, videoconferencing, and videophony, as well as television, sound, and distribution services. For these real-time services, if a cell is corrupt or lost, retransmission of cells is not possible due to the additional delay of waiting for the retransmitted cell.

The main advantage claimed for ATM is that the network switches are no longer bit-rate and service specific. Instead, the B-ISDN expects to carry all

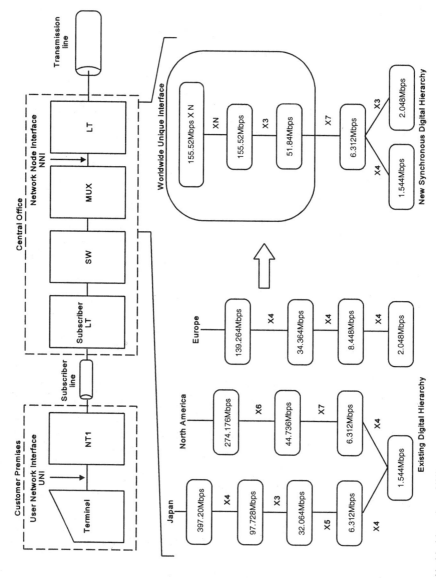

Figure 9.1 Worldwide unique NNI.

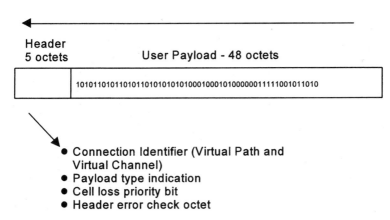

**Header
5 octets**   **User Payload - 48 octets**

1010110101101011010101010100010001010000001111001011010

• **Connection Identifier (Virtual Path and
Virtual Channel)**
• **Payload type indication**
• **Cell loss priority bit**
• **Header error check octet**

Figure 9.2 ATM cell.

services, including future and as yet unspecified services. A common user-network interface exists for all services. Within this framework, the user may request a variety of parameters that define the QoS required by the user. Examples of these parameters for an ATM-based network are bounds on the cell delay variation or cell loss. These effects are not experienced in synchronous networks and have an impact on the design of multimedia systems for the B-ISDN.

An ATM-based network will, in principle, provide the user with whatever bit rate is required (within the constraints of the interface and the network) so that teleconference users, for example, could choose the optimum picture quality required by sessions. Additionally, new television services at different bit rates could be transmitted over the network through the same user-network interface. There continues to be an improvement in picture coding algorithms and advances in technology that allow more complex algorithms to be implemented. Multimedia terminals can take advantage of service providers that offer either an improved QoS at the same average bit rate, or the same QoS at a lower average bit rate. An ATM-based network has the flexibility to provide additional transmission capacity, when required, and allows for the development of a variable bit rate and constant quality coding scheme.

An ATM connection consists of a concatenation of ATM layer links. The ATM layer does not process the user payload. Figure 9.3 shows an example of an ATM connection in terms of the B-ISDN protocol reference model. Service-specific functions reside above the solid line at the network edge.

The basis of an ATM connection is the virtual channel (VC) which can be viewed as a cell pipe. Cells arrive at the receiver in the order in which they were transmitted but with some variation in the delay. ATM errors are such that some cells may be missing due to congestion in the network, and some may be

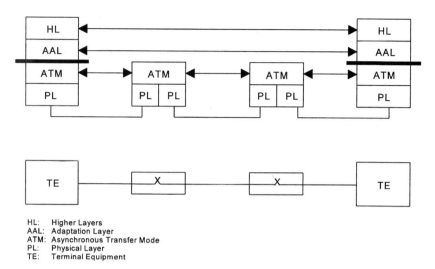

HL: Higher Layers
AAL: Adaptation Layer
ATM: Asynchronous Transfer Mode
PL: Physical Layer
TE: Terminal Equipment

Figure 9.3 B-ISDN layered protocol communication.

incorrectly sent to the receiver due to bit errors in the ATM cell header. In addition, the user data in some cells may contain errors.

ATM operates in a connection-oriented mode. At connection setup, the user of the ATM layer negotiates a traffic contract with the network. The network may allocate resources at the start of the call to meet the required QoS. The network may not accept the connection if the QoS of existing connections cannot be maintained. Throughout the connection duration, the network polices the user traffic so that the traffic contract is not violated. Network resources are released at the end of the call.

The ATM adaptation layer (AAL) enhances the ATM layer to support the functions required by the next higher layer in a service-specific manner. AAL functions typically include:

- Error control of transmission errors in the information field;
- Segmentation and reassembly of higher layer information into ATM cells;
- Handling of cell loss and misinserted cells in the ATM layer;
- Handling of cell delay jitter;
- Transfer of timing information.

To minimize the proliferation of AAL protocols, four service classes are recognized, based upon the end-to-end timing relationship, constant or variable bit rate, and connection mode. Class A and Class B cover real-time services.

The former is for constant bit-rate services, while the latter is for variable bit-rate services.

Particular AAL types under consideration for audiovisual or multimedia communications are types 1, 2, and 5.

9.1.1 End-to-End Timing

In real-time multimedia systems it is often important that the clock of the receiver have a well-defined relationship with that of the transmitter. If the receiver is running fast, it may run out of data to present to the codec; however, if it is running slow, the receive buffer may overflow. In the N-ISDN, as well as in most synchronous networks, the bit timing of the transmitter and receiver is derived from the high quality network clock. This clock is propagated by the network, thus giving the multimedia terminal a clock that may be used to present bits to the codec.

In the ATM network, the bit timing at the interface does not necessarily correspond to the timing within the virtual channel between the endpoints. In order to recover the end-to-end timing, two methods are used. The first method is through the use of a residual timestamp contained in the AAL-1 cell. In this method, called synchronous residual time stamp (SRTS), information about the difference between a derived clock and the SDH network clock is passed to the remote receiver. The receiver uses this information as an error signal in a phase locked-loop that generates its service clock.

The second way to obtain end-to-end timing is the use of the adaptive clock recovery method. This method consists of making small adjustments to the speed of the receiver's clock when the endpoint detects that it is ahead or behind the transmitter's clock. The exact method used to determine the clock adjustment depends on the application; however, one possible method is to use the fill level of the receiver's buffer. For this method to work well under most conditions, one of the endpoints must have an independent clock while the other performs the clock recovery. This method of clock recovery can be used in conjunction with any AAL type, but is the only choice for AAL-5.

9.1.2 Constant Bit Rate and Variable Bit-Rate Coding

Restrictions of traditional circuit-switched networks have meant that all commercial digital video codecs operate at a constant bit rate (CBR), despite the inherently varying information content of a motion video sequence (being dependent on, for example, changing image complexity, degree of motion, and frequency of scene changes). The internally varying rate in these codecs is smoothed by buffering and dynamic control of codec parameters (such as frame rate and quantizer stepsize) to ensure that the buffer neither empties nor overflows. Such codecs operate at a fixed bit rate, but with variable quality.

ATM networks support variable bit rate (VBR) coded video, allowing the transmitted bit rate to dynamically reflect the information content of the changing video signal, limited by the maximum channel capacity and parameters agreed upon with the network management system. A VBR codec can therefore (usually) maintain a fixed-quality, VBR mode of operation.

9.2   Adaptation of H.320 Visual Telephone Terminals to B-ISDN Environments (H.321)

Recommendation H.321 describes the technical specifications for adapting narrowband visual telephone terminals, as defined in Recommendation H.320, to B-ISDN environments. The terminal, conforming to this recommendation, interworks with the same type of terminals (that is, other H.321 terminals) accommodated in B-ISDN as well as existing H.320 terminals accommodated in N-ISDN. The generic architecture of an H.321 terminal is shown in Figure 9.4, where the constituent elements with the corresponding recommendations are indicated.

Some of the functionality supported by H.321 terminals is also supported by broadband audiovisual terminals defined in Recommendation H.310 (see Section 9.3). The interworking between H.310, H.321, and H.320 terminals is a mandatory requirement. Interworking between H.320 and H.321 terminals is achieved since the different H.321 terminals include the same functions supported by the corresponding H.320 terminals. Interworking between H.320/H.321 and H.310 terminals is achieved through a common set of H.320/H.321 functions (defined in Recommendation H.310). For example, in addition to supporting the ITU-T Recommendation H.262 video (MPEG2 video), H.310 terminals shall support Recommendation H.261, which is part of both H.320 and H.321.

9.2.1   Terminal Capabilities

With respect to the mandatory audio and video capabilities, an H.321 terminal must conform to the requirements of Recommendation H.320. This means that an H.321 terminal must support H.261 at QCIF resolution for video coding as well as G.711 for audio coding. In addition, most terminals support H.261 in CIF mode and G.728 or G.722 for audio coding. The optional H.263 video codec, which was designed for the lower bit-rate terminals, is also a part of some H.320/H.321 terminals as it offers higher quality for many bit

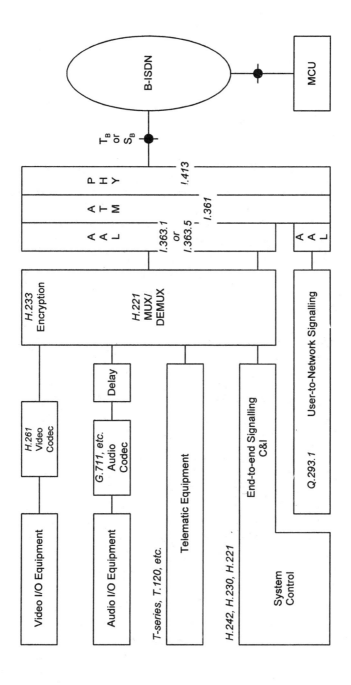

Figure 9.4 H.321 scope.

rates. H.263 also allows for higher resolutions, such as 4CIF and 16CIF, which may be attractive for H.321 terminals running at the higher bit rates.

9.2.2   AAL-1

In H.321 terminals, two methods of adapting the H.320 functions over the B-ISDN are defined. The first is achieved through the use of AAL-1. Both segmentation-and-reassembly (SAR) and convergence sublayer (CS) functions, as defined in Recommendation I.363, are considered in this recommendation. The CS layer in AAL-1 receives a constant bit-rate stream of octets from the H.221 multiplex and forms PDUs for the SAR layer. For H.321, this is done using the structured data transfer (SDT) method. In this method, octets are grouped and aligned with one or more cells according to the bearer mode. For example, when the two H.321 endpoints agree to connect at the H_0 rate (384 Kbps), a multiple of 6 octets is aligned with a pair of cells and an SDT pointer is placed at the beginning of the first cell. Since the SAR layer needs one octet, this allows for 46 octets in the first cell and up to 47 in the second. The SDT pointer indicates which octet is the last valid octet in the pair; for example, 90 would indicate that the 44th user data octet in the second cell is the last valid octet.

The SAR layer receives 47-octet PDUs from the CS layer, adds a single octet containing a sequence number, and places the resulting 48 octets into the payload of one ATM cell. The receiver can use the sequence number and knowledge of the required transfer rate to determine if a cell has been lost and generate the necessary number of "fill octets" to keep the bit rate constant.

9.2.3   AAL-5

The second method of adapting the H.320 functions over the B-ISDN is through the use of AAL-5. In this method there is the addition of a network adaptation layer (NAL) which adapts the H.320 information to the packet-oriented AAL-5. The NAL is responsible for aligning the octets of the H.221 frames into AAL-5 PDUs according to the number of B channels negotiated in the call setup. For example, in the case of the H_0 rate, each AAL-5 PDU must be a multiple of 6 octets. The NAL is also responsible for reproducing the necessary clocks for the transmitter and the receiver, using the procedures described in Section 9.2.3, if appropriate. In the case of errors, the NAL uses the 32-bit CRC within the AAL-5 PDU to determine if an error has occurred. The exact method of handling errors is implementation dependent, and could be as simple as generating "fill" bits to keep the bit rate constant. Other methods that are more complex may also be used as long as the correct clocking is maintained.

9.2.4   Adaptive Clocking

Since H.321 terminals are based on a constant bit-rate stream, the end-to-end timing relationship of the transmitter and the receiver must be well defined. If both of the endpoints have access to the N-ISDN network clock, they will remain synchronized due to the clock propagation on the N-ISDN. If one or both of the terminals are without the clock reference, the adaptive clock recovery method is used. During the call setup phase, the two terminals negotiate which of them will be performing the adaptive clock recovery. If one endpoint has a network clock reference (for example, in an H.320 to H.321 gateway), the other terminal will adjust its clock. However, if neither terminal has a network clock reference, one of the terminals must generate an independent clock from which the other will derive its clock.

9.2.5   Packetization Delay and Overhead

Delay is always an important aspect of multimedia communication systems, especially for real-time, two-way interactive videoconferencing. In H.320 systems, delay was minimized through the use of 10-ms frames within the H.221 multiplex which could be placed directly into the ISDN. Because of this fixed frame size, even as the ISDN rate increases, the multiplex delay does not decrease. In the case of H.321 systems, however, the H.221 multiplex frames cannot be placed directly into the ATM VC, but must be packetized according to the adaptation layer being used. In the case of AAL-1, delay is bounded to no more than 2 cells, which would be about 2 ms at the H_0 rate. For AAL-5, delay is not bounded due to the flexibility of the PDU size; in addition, there is a trade-off between overhead and delay due to the 8-octet trailer. It is important to note, however, that since AAL-5 does not use sequence numbers the packetization delay must be greater than the cell delay variation. This variation depends on the QoS requested for the particular VC. Recommendation H.321 gives some example packetization delays which may be used by some implementations. For an H.321 terminal connected at the H_0 rate, a 6-cell AAL-5 PDU would give a packetization delay of about 6 ms.

    The overhead incurred by the use of the AAL packetization schemes directly impacts the rate at which the two H.321 systems must connect in order to insure that the multimedia data is delivered on time. The overhead is the ratio of the number of octets that must be used for headers or trailers to the number of octets of multimedia data. The actual connection rate requested for the ATM virtual channel must be that much greater than the multimedia rate in order to fit the overhead into the channel. In the AAL-1 H_0 example given earlier, the actual connect rate would need to be at least 410 Kbps in order to account for the 7% overhead. For the 6-cell AAL-5 case, the actual connected rate would need to be at least 401 Kbps to account for the 4% overhead. Note that the 5 octets of the ATM layer are not included in the overhead calculations as these are handled by the network.

9.2.6   AAL-1/AAL-5 Interworking

Interworking between H.321 terminals that do not support the same AAL type is achieved through the use of an interworking unit (IWU). Negotiation for the use of an AAL-1 to AAL-5 IWU is performed during the initial call setup. If the H.321 terminal receiving the call determines that an IWU is necessary, it signals the address of the IWU when it rejects the call. The calling terminal places a call to the IWU with the destination terminal's address in the generic information transfer (GIT) element, and the IWU forwards the call to the destination terminal.

9.2.7   Applications for H.321

One of the motivations for the use of H.321 and ATM in place of H.320 over N-ISDN is the lower cost of conferences at the higher bit rates. Since ATM is designed as a network that can handle data, voice, and video, a corporation can use ATM as its network backbone and share access to an N-ISDN or B-ISDN WAN interface. Also, an ATM switch is usually cheaper than an N-ISDN switch handling the same bit rates.

For desktop videoconferencing (DVC) systems (H.321 endpoints running on desktop computers), the ATM interface can double as the data network interface. This enables data applications such as Internet web browsing or database access along with the H.321 conferencing. By choosing a single ATM interface into the computer instead of a combination of Ethernet and N-ISDN, a corporation can lower the cost of DVC installation and maintenance. In addition, other ATM related services (such as H.310) can be incorporated into the overall corporate architecture.

9.3   High-Resolution Broadband Audiovisual Communication Systems (H.310)

The resolution of the H.321 is generally limited to the CIF quality specified in the H.261 standard (although H.263 may also be supported by an H.321 terminal to obtain higher resolutions). The purpose of the H.310 terminal is to provide a higher resolution (full frame rather than one field) capability as defined by the MPEG2/H.262 video coding standard.

ITU-T Recommendation H.310 terminals are intended for the support of the following applications via the B-ISDN network:

• Conversational services (videoconferencing and videotelephony);
• Retrieval services;
• Messaging services;
• Distribution services with user individual presentation control (video-on-demand);
• Distribution services without user individual presentation control (broadcast TV services);
• Video transmission or surveillance.

Figure 9.5 shows a generic H.310 broadband audiovisual communication system.

9.3.1 Communication Mode

A bidirectional audiovisual terminal is called a receive-and-send terminal (RAST) while a unidirectional terminal may be a receive-only terminal (ROT) or a send-only terminal (SOT). Each of these H.310 terminal types can run over AAL-1 or AAL-5, making for six combinations (RAST-1, RAST-5, ROT-1, ROT-5, SOT-1, SOT-5) as well as three mixed types (RAST-1/5, ROT-1/5, SOT-1/5).

The capabilities of H.310 terminals are classified into five attributes: video codec capabilities (VCC), audio codec capabilities (ACC), network adaptation capabilities (NAC), control and indication capabilities (CIC), and other data capabilities (ODC). A communication mode is defined as a combination of those capabilities that are employed by a terminal at an instance of audiovisual communication. Since a communication session may be asymmetric for many applications in the broadband environment, these attributes are specified separately for the transmit end (TE) and the receive end (RE).

9.3.2 Video Codec Capabilities

Both H.261 and H.262 video capabilities are considered in this recommendation. H.262 is the ITU-T Recommendation that was developed jointly with MPEG2 (see Chapter 11). All H.310 RAST terminals support ITU-T Recommendation H.261 with both the CIF and QCIF picture resolutions. This enables the interworking between H.310 and a wide range of existing and future H.320/H.321 terminals. In addition, all H.310 terminals must be capable of

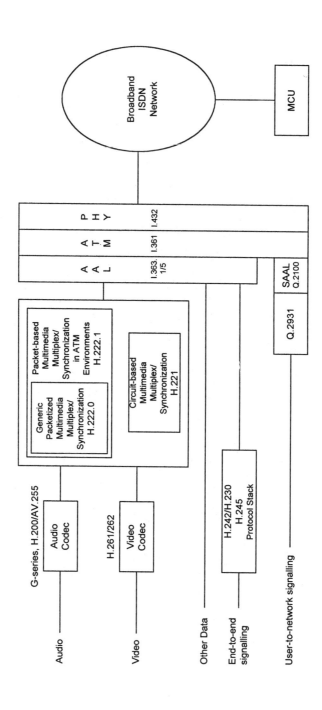

Figure 9.5 H.310 scope.

decoding any coded video bitstream that is in conformance with Recommendation H.262 Main Profile (MP) at Main Level (ML). H.262 MP at High Level (MP@HL) and MP at High-1440 Level (MP@H14L) are supported by some H.310 terminals. Therefore in addition to supporting the H.261 (CIF and QCIF) video coding modes, all H.310 RAST terminals support at least one of the following video standards:

- H.262 MP@ML;
- H.262 MP@H14L;
- H.262 MP@HL.

Recommendation H.263 can also be supported by H.310 terminals, although it is not required for interworking with the narrowband or H.321 terminals.

9.3.3 Audio Codec Capabilities

Three types of audio coding standards are considered in Recommendation H.310: ITU-T G-series (G.711, G.722, and G.728), ISO/IEC 11172-3 (MPEG1 audio), and ISO/IEC 13818-3 (MPEG2 audio). All RAST H.310 terminals must support ITU-T Recommendation G.711. This enables existing and future H.320/H.321 terminals to interwork with H.310 terminals. In addition, all H.310 terminals should be capable of decoding any audio bitstream that is in conformance with the ISO/IEC 11172-3 (MPEG1 audio) standard. The support of the ISO/IEC 13818-3 audio standard (MPEG2 audio) as well as G.722 and G.728 coding is optional for H.310 terminals.

9.3.4 Network Adaptation Capabilities

The network adaptation capabilities of H.310 terminals include the multimedia multiplex and synchronization mechanism, ATM adaptation layer, transfer rate, and ATM VC capabilities. Multiplexing and synchronization of audio and video can be accomplished in two ways, using either the program stream (PS) or the transport stream (TS) of Recommendation H.222.0 (also called ISO/IEC 13818-1). The program stream is designed for systems where there is little or no loss in the communication channel. This mode is optional for H.310 terminals, but may be requested when the QoS is sufficient. The transport stream is designed for systems where there is a possibility of errors or loss in the channel. Support for this mode is required by Recommendation H.310. In addition to providing a clock recovery and error control mechanisms, the transport stream allows for multiplexing of multiple program streams, making possible content selection by the remote terminal.

H.310 requires that endpoints support two transfer rates, 6.1 and 9.2 Mbps, that are transferred using either AAL-1 or AAL-5, depending on the terminal type. The actual determination of these parameters is done by the end-to-end signaling element, defined by Recommendation H.245.

9.3.5   End-to-End Signaling

H.310 uses Recommendation H.245 as the control mechanism to signal capabilities and modes between the endpoints. H.245 is carried over an AAL-5 VC that is established at the beginning of an H.310 session and is used to control the session. H.245 has the concept of logical channels, which represent the channels of communication between the endpoints. In the case of H.310, opening of a logical channel corresponds to the opening of a virtual channel through which the media will be exchanged. This virtual channel carries either the program stream or transport stream of H.222.0. The parameters for the selection of the stream type are included in the channel establishment messages.

The H.245 channel can be used to request various modes of operation, such as switching the video resolution or frame rate. Since the capabilities of the remote terminal are exchanged at the outset, each terminal knows what modes are possible and can present the correct options to the user. In addition to changing the audio and video modes, H.245 has procedures for lower-level functions like determination of the round trip delay and performing diagnostic loop-backs. In all, H.245 provides a flexible control mechanism, which is why it was adopted for the H.324 and H.323 Recommendations.

In addition to the H.245 end-to-end signaling, H.310 supports the optional use of digital storage media—command and control (DSM-CC) signaling defined by ISO/IEC 13818-6. DSM-CC was designed for distribution services such as video-on-demand and can augment the H.310 system with a more flexible mechanism for selecting video programming. DSM-CC can also be used to control messaging and retrieval services within the H.310 framework.

9.3.6   Packetization Delay and Overhead

As in H.321 systems, delay is an important aspect of H.310 systems, especially for real-time, two-way interactive videoconferencing (RAST terminals). Unlike H.321, however, H.310 systems have many more sources of overhead and delay. First, the transport stream of H.222.0 is constructed from packetized elementary stream (PES) packets from the audio and the video encoders. The PES packets already contain a small overhead in the form of a header that contains flags, timestamps, and stream information. Multiple PES packets from the audio and video streams may be placed into the 188 octet TS packet, which contains additional header information. Finally, for AAL-5, one or more TS packets are placed into an AAL-5 PDU. In order to keep the packetization delay low, the system must produce a small number of TS packets per AAL-5 PDU; as such, all H.310 systems must support receiving only two TS packets per PDU.

The video subsystem within H.310 can be an additional source of delay. Recommendation H.262 normally specifies the use of bidirectional-interpolated frames (B-frames) within the video syntax along with the intra-frames (I-frames) and predictive frames (P-frames) used in H.261. B-frames are constructed using information from a future frame and a previous frame as shown in Figure 9.6. Since a B-frame can neither be encoded nor decoded without the information from the future frame, there is a delay introduced that is equal to the time between two unidirectional frames. Taking the example of two B-frames inserted for each I- or P-frame, and assuming a rate of 30 frames per second, an additional round-trip delay of at least 133 ms is introduced when using B-frames. Due to this large additional delay, implementations of H.310 used for conversational services often choose to encode video with only the I- and P-frames. These frames can be decoded and presented without waiting for future frames to arrive.

9.3.7 Intercommunication With N-ISDN Terminals

It is desirable that RAST H.310 terminals be able to interact with H.320/H.321 terminals. In an intercommunication session with an H.320/H.321 terminal, RAST H.310 terminals function as H.321 terminals. RAST-1 terminals must support H.321 using AAL-1, while RAST-5 terminals must support H.321 using AAL-5. Support of B, 2B, and H_0 communication modes is mandatory, while other communication modes are optional. Since H.310 incorporates H.321 functionality, there is no need for a gateway between H.310 and H.321

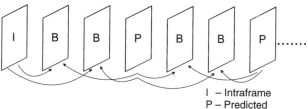

I – Intraframe
P – Predicted
B – Bidirectional Interpolation

Figure 9.6 H.262 video B-frames.

terminals on the B-ISDN. In addition, H.310 RAST terminals can use the same N-ISDN interworking unit used by H.321 terminals.

9.3.8 Applications for H.310

Since H.310 defines a wide range of terminal capabilities and options, conforming terminals have applications in many different areas beyond the conversational services of videoconferencing or videophony. Since MPEG2 is the dominant standard for the broadcast TV industry there are many requirements for the communication of these signals with very high quality. A combination of a single SOT and multiple ROT terminals can implement a broadcast-quality distribution network that contains multiple audio and video streams. Since H.222.0 allows for multiple program streams within the transport, individual ROT terminals can select programming either automatically or under user control. This can be implemented as a video-on-demand system or a broadcast TV service. A video surveillance system could be implemented with multiple SOT terminals and a single ROT that can select the camera view desired.

Unlike H.321, H.310 systems have a much higher cost for conversational services when compared to H.320 on N-ISDN, but provide a much higher quality. Even when a corporation has the ATM infrastructure and B-ISDN connectivity, it may choose to use H.321 for standard conversational services (such as DVC systems) and H.310 for higher quality services (such as in large conference rooms). Since H.310 RAST terminals must support communication with H.321 terminals, a large mixed environment is possible.

10

Multipoint Graphic Communications (T.120)

In many situations it is not necessary to transmit a motion video signal to achieve a satisfactory electronic conferencing capability. Instead, the conference may require only audio and graphics. This chapter describes a new ITU standard, T.120, which provides the ability to interactively exchange graphics (or any data) on a multipoint basis. The audio portion of the conference would be provided independently from the T.120 process.

Traditionally, telephony services have been confined to point-to-point operation. To support group activities, such as meetings and conferences involving physically separated participants, there is a requirement to join together more than two locations. The term *multipoint communication* simply describes the interconnection of multiple terminals as shown in Figure 10.1. Normally, a special network element, known as an MCU, or simply a bridge, is required to provide this function. The raw communication data stream would typically consist of one or more of the three media elements: audio, video, and data. The term *multimedia* is now widely used to collectively describe this grouping.

The T.120 series of recommendations applies to the data element, which would typically be used both to provide a data communications service and management of any other media services present. The T.120 ITU Recommendation introduces the T.12x series of standards, collectively referred to as the T.120 series. The T.120 protocol is a means of telecommunicating all forms of data/telematic media between two or more multimedia terminals and of managing such communication. It can also manage real-time conversational speech and video whose information signals are transmitted on channels separate from that carrying the T.120 protocol.

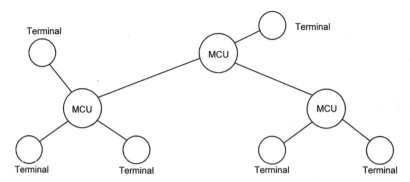

Figure 10.1 Typical multipoint configuration.

The T.120 model is composed of a communications infrastructure and application protocols that make use of it. Figure 10.2 shows the full model with both standardized and nonstandardized components. The model serves both to show the scope of the T.120 suite of recommendations and the relationship between each of the recommendations and other elements in the system. Generally, each layer provides services to the layer above and communicates to its peer(s) by sending protocol data units (PDUs) via services provided by the layer below. This discussion will address each of the major functional levels in Figure 10.2, that is, application protocols, node controller, and communications infrastructure.

10.1 Node Controller

In a T.120-based system, the term *node controller* is used to describe the management function or role at a terminal or MCU. It is the node controller that issues the primitives to generic conference control (GCC) that start and control the communication session. The node control function will exist in all systems, providing the interface above GCC and performing any other required management functions. The node controller itself is outside the scope of the T.120 recommendations, and only where it communicates to GCC and any other standardized elements are the interfaces defined (by association).

10.2 Communications Infrastructure

The communications infrastructure, illustrated in Figure 10.2, provides simultaneous multipoint connectivity with reliable data delivery. It can accommodate multiple independent applications concurrently using the same multipoint

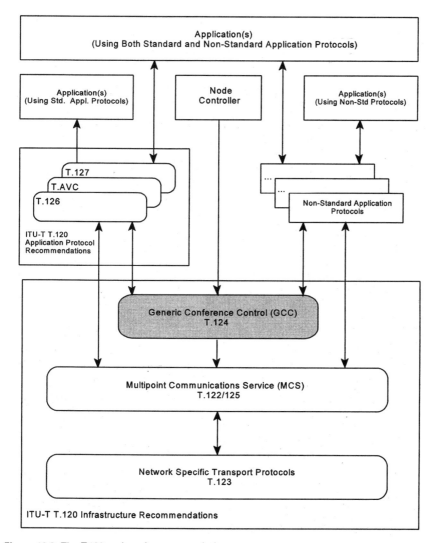

Figure 10.2 The T.120 series of recommendations.

environment. Connections can be any combination of circuit-switched tele-communications networks and packet-based LANs and data networks. It is composed of three standardized components: generic conference control, multipoint communications service (MCS), and transport protocol profiles for each of the supported networks. Adding GCC provides a range of facilities oriented to the use of MCS in an electronic meeting or conference but believed to be generally useful for other multipoint communications requirements.

10.2.1 T.122/125 Multipoint Communications Service

T.122 defines the multipoint services available to the developer, while T.125 specifies the data transmission protocol. Together they form MCS, the multipoint "engine" of the T.120 conference.

MCS relies on T.123 to actually deliver the data. (Use of MCS is entirely independent of the actual T.123 transport stack(s) that is loaded.) MCS is a powerful tool that can be used to solve virtually any multipoint application design requirement. MCS is an abstraction of a rather complex organism. Learning to use MCS effectively is the key to successfully developing real-time applications.

10.2.1.1 How MCS Works

In a conference, multiple endpoints (or MCS nodes) are logically connected to form what T.120 refers to as a domain. Domains generally equate to the concept of a conference. Applications may actually be attached to multiple domains simultaneously. For example, the chairperson of a large online conference may simultaneously monitor information being discussed among several activity groups. If the chairperson wanted to bring two of these groups together to share ideas, the conference provider could use the sophisticated domain merge facility to accomplish this request.

In a T.120 conference, nodes connect upward to an MCU. The MCU model in T.120 provides a reliable approach that works well in both public and private networks. Multiple MCUs may be easily connected in a single domain. Figure 10.3 illustrates potential topology structures. Each domain has a single top provider or MCU that houses the information base that is critical to the conference. If the top provider either fails or leaves a conference, the conference is terminated. If a lower level MCU (that is, not the top provider) fails, only the nodes on the tree below that MCU are dropped from the conference.

One of the critical features of the T.120 approach is the ability to direct data. This capability allows applications to communicate very efficiently. MCS applications direct data within a domain via the use of channels. While an application may only send data along one channel at a time, it may simultaneously subscribe or receive information from many channels at once. These channel assignments can be dynamically changed during the life of the conference.

It is up to the application designer to determine how to use channels within an application. For example, an application may send control information along a single channel and application data along a series of channels that may vary depending upon the type of data being sent. The application developer may also take advantage of the MCS concept of private channels to direct data to a discrete subset of a given conference.

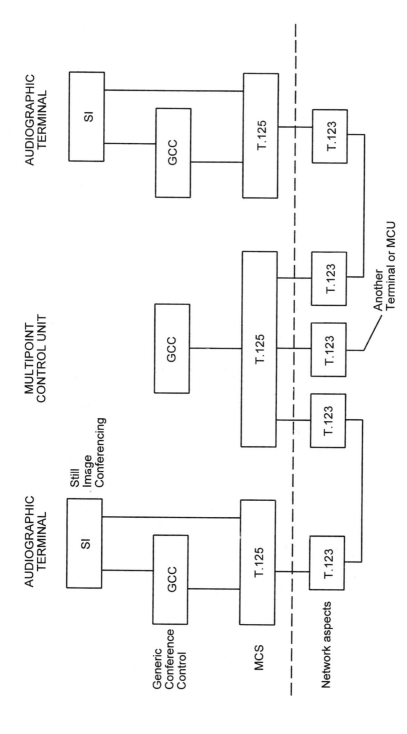

Figure 10.3 Framework of the AGC protocol suite.

Data may be sent with one of four priority levels. MCS applications may also specify that data are routed along the quickest path of delivery using the standard send command. If the application uses the uniform send command, it ensures that data from multiple senders will arrive at all destinations in the same order. Uniform data always travel all the way up the tree to the top provider.

There are no constraints on the size of the data sent from an application to MCS. Segmentation of data is automatically performed on behalf of the application. However, it is the application's responsibility to reassemble the data upon receiving them by monitoring flags provided when the data are delivered.

Tokens are the last major facility provided by MCS. Services are provided to grab, pass, inhibit, release, and query tokens. Token resources may be used as either exclusive (that is, locking) or nonexclusive entities.

Tokens can be used by an application in a number of ways. For example, an application may specify that only the holder of a specific token (that is, the conductor) may send information in the conference. Another popular use of tokens is to coordinate tasks within a domain. For example, suppose a teacher wants to be sure that every student in a distance learning session answered a particular question before displaying the answer. Each node in the underlying application inhibits a specific token after receiving the request to answer the question. The token is released by each node when an answer is provided. In the background, the teacher's application continuously polls the state of the token. When all nodes have released the token, the application presents the teacher with a visual cue that the class is ready for the answer.

10.2.2   T.124 Generic Conference Control

GCC provides a comprehensive set of facilities for establishing and managing the multipoint conference. It is with GCC that we first see features that are specific to the electronic meeting.

At the heart of GCC is an important information base about the state of the various conferences it may be servicing. One node, which may be the MCU itself, serves as the top provider for GCC information. Any actions or requests from lower GCC nodes ultimately filter up to this top provider.

Using mechanisms in GCC, applications create conferences, join conferences, and invite others to conferences. As endpoints join and leave conferences, the information base in GCC is updated and can be used to automatically notify all endpoints when these actions occur. GCC also knows who is the top provider for the conference. However, GCC does not contain detailed topology information about the means by which nodes from lower branches are connected to the conference.

Every application in a conference must register its unique application key with GCC. This enables any subsequent joining nodes to find compatible applications. Furthermore, GCC provides robust facilities for applications to exchange capabilities and arbitrary feature sets. In this way, applications from different vendors can readily establish whether or not they can interoperate and at what feature level. This arbitration facility is the mechanism used to ensure backward compatibility between different versions of the same application.

GCC also provides conferences security. This allows applications to incorporate password protection or lock facilities to prevent uninvited users from joining a conference.

Another key function of GCC is its ability to dynamically track MCS resources. Since multiple applications can use MCS at the same time, applications rely on GCC to prevent conflicts for MCS resources, such as channels and tokens. This ensures that applications do not step on each other by attaching to the same channel or requesting a token already in use by another application.

Finally, GCC provides some capabilities for supporting the concept of conductorship in a conference. GCC provides applications with information about who the meeting conductor is, and a means to transfer the conductor's "baton." The developer is free to decide how to use these conductorship facilities within the application.

10.2.3 T.123 Transport Protocol Stack Profiles

MCS expects its underlying transport to provide reliable point-to-point data delivery of its PDUs, and to segment and sequence that data if necessary. T.123 (Figure 10.4) is designed to provide open and easily extended network support for both standardized and nonstandardized protocols. The basic T.123 presents a uniform OSI transport interface and services (X.214/224) to the MCS layer above. Connection-oriented profiles are provided for switched telecom and packet-switched networks. For computer networks that are not connection-oriented, the profiles include additional transport layer capability to make them functionally connection-oriented.

The T.120 suite provides for operation over the following networks:

- PSTN or compatible service with 3.1-kHz bandwidth channels using V series modems;
- ISDN as defined in ITU-T I series recommendations;
- Packet-switched data network (PSDN) using X.25;

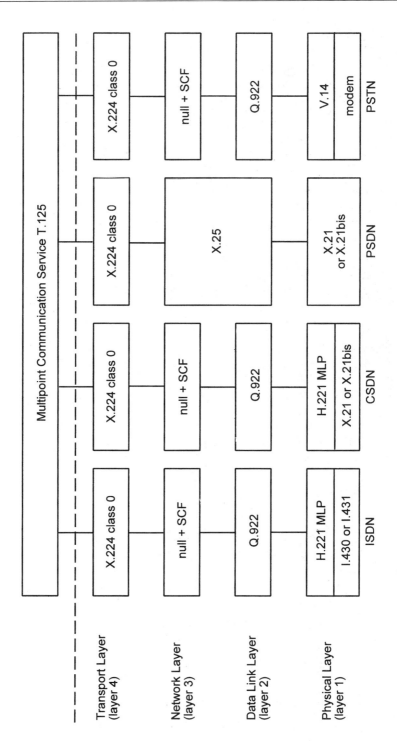

Figure 10.4 Basic mode profiles general structure.

- Other (switched or permanent) digital circuits using H.221 framed signals;
- Use of T.120 on other networks such as LANs and ATM is currently under study.

10.3 Application Protocols

Application protocols comprise a set of PDUs and associated actions for application peer-to-peer(s) communication. These may be proprietary protocols, or they may be standardized by the ITU-T or other international or national standards bodies. The T.120 series includes a set of application protocols designed to support common facilities for multipoint communication. These protocols define and mandate a minimum requirement to ensure interworking between different implementations. They also include facilities for simultaneous multipoint file transfer (T.127) and audiographics protocols for still image viewing and annotation, and application sharing and fax (all provided in T.126).

10.3.1 Recommendation T.126: Still Image Exchange and Annotation (SI)

T.126 defines a protocol for viewing and annotating still images transmitted between two or more applications. This capability is often referred to as document conferencing or shared whiteboarding.

An important benefit of T.126 is that it readily shares visual information between applications that are running on dramatically different platforms. For example, a Windows-based desktop application could easily interoperate with a collaboration program running on a Power Mac. Similarly, a group-oriented conferencing system, without a PC-style interface, could share data with multiple users running common PC desktop software.

T.126 presents the concept of shared virtual workspaces that are manipulated by the endpoint applications. Each workspace may contain a collection of objects that include bitmap images and annotation primitives, such as rectangles and freehand lines. Bitmaps typically originate from application information, such as a word processing document or a presentation slide. Because of their size, bitmaps are often compressed to improve performance over lower-speed communication links.

T.126 is designed to provide a minimum set of capabilities required to share information between disparate applications. Several important concepts commonly found in shared whiteboarding applications, such as support for nonbitmap data (objects), are not supported in T.126. Additionally, because T.126 is simply a protocol, it does not provide any of the API-level structures that allow application developers to easily incorporate shared whiteboarding into an application.

10.3.2   Recommendation T.127: Multipoint Binary File Transfer Protocol

This recommendation defines a protocol to support the interchange of binary files within an interactive conferencing or group working environment where the T.120 series of standards is in use. It provides mechanisms that facilitate distribution and retrieval of one or more files simultaneously. T.127 uses services provided by T.122 (MCS) and T.124 (GCC). T.127 adheres to the header structure defined in T.434.

10.3.3   Recommendation T.128: Application Sharing

T.128 defines a protocol for application sharing. This capability allows an application being run on a computer at the remote end of the conference to be viewed on the computer at the local end of the conference. In addition to viewing the application, this protocol also allows the local computer to control the application running on the remote computer.

The application that is being shared is not aware that it is being shared, and therefore does not require any special capability to allow sharing. The application does not need to be present on both machines, and in fact, T.128 provides cross-platform operation. This allows an application running on one type of computer (for example UNIX) to be viewed on a different type of computer (for example Microsoft Windows). The T.128 protocol provides the common link between the different computers.

T.128 has two modes of operation. A legacy mode provides compatibility with an early application sharing package from PictureTel and Microsoft called Group Share®. This mode allows a T.128 application sharing package to interoperate with a system which is using this older package. The normal mode provides similar capability within the T.120 architecture, making use of the T.120 PDU encoding rules and the GCC capability negotiation. This mode is also used as the basis for future extensions, while the legacy mode is expected to phase out over time.

In the application sharing environment, a window is established on the local computer which displays a view of the remote computer's display, which shows the application being shared. As the application display changes, the local window is updated. In addition to being able to view the remote application, the local computer can also be given control of the application. In this case, keyboard and mouse actions from the local computer are sent to the application running on the remote computer. This allows the local user to control the application as if sitting at the remote computer.

10.3.4 Recommendation T.134: Text Chat

Initially developed to provide communications capability for the hearing impaired, T.134 provides point-to-point and multipoint distribution of text messages within a T.120 conference. This provides real-time or near real-time text communications for those applications where audio communication is not available.

10.3.5 Recommendation T.135: Reservations

T.135 provides a protocol for a user to communicate with a multipoint conference reservation system. This system allows the user to reserve and control multipoint conference resources (for example MCUs). T.135 provides a mechanism to create, modify, or delete a conference. The available MCU resources can be checked to see if the desired conference can be supported at the desired time. When a conference is created, a list of sites and other information about the characteristics of the conference is entered. This list of sites can join the conference, or the MCU can invite them into the conference. Information on the capability of each site can be stored in advanced and used when establishing the connection to that site. Site information can be listed, modified, or deleted.

When the conference starts, T.135 provides mechanisms for adding new sites to the conference and extending the duration of the conference, if resources are available. A user may also get a listing of existing conferences that he or she may want to join.

10.3.6 Recommendation T.RDC: Remote Device Control

T.RDC is a new recommendation that provides control of remote audio and video devices during a conference. A similar capability will be provided by H.RDC but will not require the use of T.120.

Remote device control is an extension of the (H.281 Far End Camera Control. It provides a mechanism to select a remote-end camera source; control the pan, tilt, zoom, and focus movements of the camera; save the current camera position as a preset; and move to a previously saved preset position. Mechanisms are also provided for selecting the audio source, and controlling the source volume.

Control of other devices such as VCRs and scanners may also be provided as the recommendation matures. The management and control of network elements such as MCUs, gateways, and switching systems are under consideration. In particular, MCU control of "who you see" and "who sees you" are important functions.

11

ISO Audiovisual Standards (MPEG, JPEG, JBIG)

The International Standards Organization (ISO) has established the three working groups listed below to develop standards for coding audiovisual signals.

1. Motion Picture Experts Group (MPEG)
   ISO/IECJTCI/SC2/WG11;
2. Joint Photographic Coding Experts Group (JPEG)
   ISO/IEC/JTC1/SC2/WG8;
3. Joint Bilevel Image Group (JBIG)
   ISO/IEC/JTC1/SC2/WG9.

The status of the standards which have been, and are continuing to be, developed by these three groups is summarized in the following sections.

The MPEG work is organized in four separate projects—MPEG-1, MPEG-2, MPEG-4, and MPEG-7—which are summarized below and discussed in the following sections.

1. MPEG-1: Coding of Moving Pictures and Associated Audio for Digital Storage Media at up to About 1.5 Mbps;

   • Part 1: Systems   IS[1]   11172-1 (1993);

---

1. IS, International Standard.

223

- Part 2: Video   IS   11172-2 (1993);
- Part 3: Audio   IS   11172-3 (1993);

2. MPEG-2: Generic Coding of Moving Pictures and Associated Audio Information;

- Part 1: Systems   IS   13818-1 (1995);
- Part 2: Video   IS   13818-2 (1995);
- Part 3: Audio   IS   13818-3 (1995);

3. MPEG-4: Coding of Audio-Visual Objects;

- Part 1: Systems   FCD[2]   14496-1;
- Part 2: Video   FCD   14496-2;
- Part 3: Audio   FCD   14496-3;

4. MPEG-7: Multimedia Content Description Interface.

11.1 MPEG-1

11.1.1 System Layer (11172-1)

An ISO 11172 bitstream is constructed in two layers: the outermost layer is the system layer, and the innermost is the compression layer. The system layer provides the functions necessary for using one or more compressed data streams in a system. The video and audio parts of this specification define the compression encoding layer for audio and video data. Coding of other types of data is not defined by the specification but is supported by the system layer, providing other types of data adhere to the system constraints. The system layer supports four basic functions: the synchronization of multiple compressed streams on playback, the interleaving of multiple compressed streams into a single stream, the initialization of buffering for playback start up, and time identification.

11.1.2 Video Coding (11172-2)

The MPEG-1 video coding standard specifies the coded representation of video for digital storage media and specifies the decoding process. The representation

---

2. FCD, Final Committee Draft.

supports normal-speed forward playback as well as special functions such as random access, fast play, fast reverse play, normal-speed reverse playback, pause, and still procedures. This international standard is compatible with standard 525- and 625-line television formats and provides flexibility for use with personal computer and workstation displays.

This international standard is primarily applicable to digital storage media supporting a continuous transfer rate up to about 1.5 Mbps, such as compact disc (CD), digital audio tape (DAT), and magnetic hard disks. The storage media may be directly connected to the decoder or via communication means such as busses, LANs, or telecommunication links. This international standard is intended for noninterlaced video formats having approximately 288 lines of 352 pels and picture rates around 24 to 30 Hz. The objective of the basic MPEG-1 coding algorithm is to provide VCR picture quality that is similar to H.261, that is, 8 × 8 DCT, interframe prediction, and motion compensation.

The MPEG-1 coding scheme is very similar to that of ITU-T H.261. The major difference between the two is that MPEG-1 allows bidirectional interpolation of frames (B-frames). A video sequence is divided into groups of pictures (GOPs) as shown in Figure 11.1. There are three possible types of pictures in a GOP, specifically, I-picture, P-picture, and B-picture. Coding of I- and P-pictures is similar to the scheme used for the JPEG and H.261 standards, that is, intraframe and liner prediction interframe techniques with DCT. Coding of B-pictures is slightly different from that of the P-pictures. In P-pictures, motion prediction is obtained only from some previous frame; that is, forward prediction. However, in B-pictures, information from both previous and future frames is used for prediction, that is, bidirectional interpolation.

11.1.3 Audio Coding (11172-3)

The MPEG-1 audio coding standard specifies the coded representation of high-quality audio for storage media and the method for decoding of high-quality

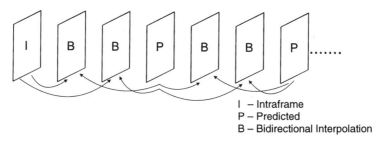

I – Intraframe
P – Predicted
B – Bidirectional Interpolation

Figure 11.1 MPEG frame coding.

audio signals. It is compatible with the current formats (CD and DAT) for audio storage and playback. This representation supports normal speed playback. This standard is intended for application to digital storage media providing a total continuous transfer rate of about 1.5 Mbps for both audio and video bitstreams, such as CD, DAT, and magnetic hard disk. The storage media may either be connected directly to the decoder or via other means such as communication lines and the MPEG systems layer. This standard is intended for sampling rates of 32, 44.1, and 48 kHz and 16-bit PCM input/output to the encoder/decoder.

11.2   MPEG-2 (13818)

11.2.1   MPEG-2 System (13818-1)

The systems part of MPEG-2 addresses the combining of one or more elementary streams of video and audio, as well as other data, into single or multiple streams that are suitable for storage or transmission. System coding is specified in two possible forms: the transport stream and the program stream. Each is optimized for a different set of applications. Both the transport stream and program stream defined in this international standard provide coding syntax that is necessary and sufficient to synchronize the decoding and presentation of the video and audio information, while ensuring that data buffers in the decoders do not overflow or underflow. Information is coded in the syntax using time stamps concerning the decoding and presentation of coded audio and visual data and time stamps concerning the delivery of the data stream itself. Both stream definitions are packet-oriented.

The basic multiplexing approach for single video and audio elementary streams is illustrated in Figure 11.2. The video and audio data are encoded as described in Parts 2 and 3 of this international standard. The resulting compressed elementary streams are packetized to produce PES packets.

The program stream is analogous and similar to the MPEG-1 systems multiplex. It results from combining one or more streams of PES packets, which have a common time base, into a single stream. The program stream definition can also be used to encode multiple audio and video elementary streams into multiple program streams, all of which have a common time base.

The transport stream combines one or more programs with one or more independent time bases into a single stream. PES packets made up of elementary streams that form a program share a common timebase. The transport stream is designed for use in environments where errors are likely, such as storage or transmission in lossy or noisy media. A simplified diagram of the packet which is used in an MPEG-2 transport stream is shown in Figure 11.3. The transport stream is composed of fixed-length 188-byte packets, always including

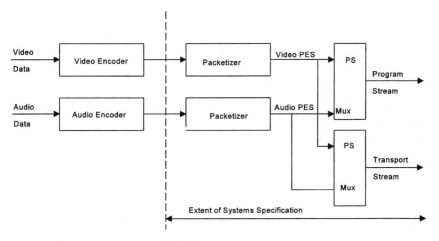

Figure 11.2 Simplified overview of MPEG-2 system.

a sync byte of a known value (hexadecimal 47). Since this value may also occur in the data, acquisition of packet boundaries normally consists of looking for the sync byte first, then verifying that it occurs 188 bytes later several times in a row. The remainder of the transport stream packet header includes a single-bit error indicator (to signal to a decoder to take concealment action, as the current packet has been damaged somewhere in transmission). The payload unit start indicator provides an easily extracted signal to a decoder that a PES packet or other payload is starting in the current transport packet; this facilitates start-up of decoding. The transport priority is a single bit indicating whether the current packet is high or low priority, although the exact choice of when a packet should be marked high or low priority and how this information should be used are not specified. The packet ID (PID) is a 13-bit address that indicates which elementary stream is carried by this packet (or, alternatively, which table type or private

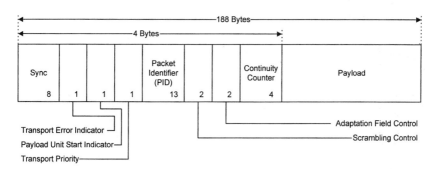

Figure 11.3 MPEG-2 transport stream packet format.

data stream is carried by this packet). The transport stream scrambling control is present (as are several other features of the transport stream) to facilitate secure transmission without actually defining the security protocol.

Occasionally, the later 184 bytes of a transport packet will include an adaptation field, whose most important content is the program clock reference. This field can also include indicators on discontinuities and on how to splice MPEG streams, private data, stuffing to adjust the transport stream rate, and a number of other parameters.

11.2.2   MPEG-2 Video (13818-2)

Like MPEG-1, MPEG-2 consists of three interrelated standards—video coding, audio coding, and system. Again like MPEG-1, MPEG-2 is a generic toolkit designed for a wide range of applications such as storage/retrieval, broadcast TV (contribution and distribution), satellite TV transmission, and high-definition TV (HDTV). In many ways, MPEG-2 is similar to MPEG-1. The primary difference is that MPEG-2 provides for the coding of an interlaced TV frame while MPEG-1 is restricted to the coding of one field. This difference, and others, are summarized in Table 11.1. The table also compares MPEG operation with H.261 coding.

11.2.2.1   Comparison of Video Coding Standards

The MPEG-2 toolkit standard is organized into levels and profiles as illustrated in Figure 11.4. In general, the level axis refers to picture resolution, where the 1440 level refers to the number of pixels per line in an HDTV picture. The profile axis describes the complexity of the set of tools required for the application. For example, the simple profile prohibits the use of B-frames. The main profile/main level (MP/ML) is the most basic and fundamental application of MPEG-2, employing a resolution and complexity appropriate for conventional broadcast TV.

Levels						
	High level		X			X
	High - 1440		X		X	X
	Main level	X	X	X		X
	Low level		X	X		
		Simple profile	Main profile	SNR scalable profile	Spatially scalable profile	High profile
		- No B-pictures - Non-scalable - 4:2:0	- Simple with B-pictures	- 4:2:0	- 4:2:0	- 4:2:2

Figure 11.4 MPEG-2 structure.

Table 11.1
Comparison of Video Coding Standards

	H.261	H.263	MPEG-1	MPEG-2 *
General standard structure	Narrow profile	Narrow profile		Generic tool kit
Picture format	176×144 (mandatory), 352×288 (optional)	SQCIF, QCIF, CIF, 4CIF, 16CIF	One field	Field or frame
Quantization precision of motion vectors	One pixel	One half pixel	One half or one pixel	
B-frames/PB frame	None	PB frames	B frames—An available tool	
Intraframe coding	Usually distributed	Flexible	Full frame is mandatory	
Color coding (Figure 11.5)	4:2:0	4:2:0	4:4:4, 4:2:2, 4:2:0	
Picture structure	GOB	GOB	Slice	
Dual prime (a special motion compensation mode)	No	No	No	Yes
Nominal bitrate	56 to 1,936 Kbps	Low bit rate	1.5 Mbps	4 to 20 Mbps
Applications	Interactive audiovisual services	VTC/videophone		- Broadcast TV + contribution + distribution
	- Videophone - Videoteleconferencing	via PSTN/mobile network	- VCR	- DBS - HDTV

* H.262 is a particular profile of MPEG-2.

MPEG-2 MP (main profile) frame-picture coding allows three motion compensation modes (that is, frame, field, and dual prime) to be adaptively selected for each macroblock in a P- or B-picture. The frame motion compensation mode is the same as the prediction mode in MPEG-1 where the odd and even lines of a macroblock are predicted as one unit using one motion vector. Field motion compensation allows the odd and even lines (fields) of a macroblock to be independently predicted using two separate motion vectors. Dual prime is a field-based prediction that allows interpolation of two reference fields using one motion vector and a correction vector. This mode is only allowed in a sequence with no B-pictures to reduce memory bandwidth requirement. MPEG-2 also allows DCT to be performed on a frame-format block or a field-format block. The latter is created by separating the odd and even lines of a macroblock. These motion prediction and frame/field DCT options allow MPEG-2 to handle interlaced video coding efficiently, which is not possible with MPEG-1. One of the MPEG-2 tools, which is changed in different profiles, is color resolution as illustrated in Figure 11.5. The most common color resolution is 4:2:0, in which case the color resolution is half the luminance resolution in both the horizontal and vertical directions. In the 4:2:2 case, the color resolution is half in only the horizontal direction. The luminance and color resolutions are the same in the 4:4:4 condition.

Scalability is an important tool which is employed in a different way in the various profiles. The general concept of scalable coding is illustrated in Figure 11.6. The general idea is to divide the MPEG-2 signal into two parts: a base layer and an enhancement layer. If a low cost, low-complexity decoder was to receive the signal, only the low resolution base-layer signal would be displayed. A fully capable decoder would add the enhancement signal to the base-layer bit stream thereby displaying the maximum resolution image.

11.3 MPEG-4 (Coding of Audiovisual Objects)

In the past, standardized coding schemes such as H.261, MPEG-1, and MPEG-2 each have defined a bit-precise syntax for efficiency and ease of decoding. These schemes also have been able to include a limited degree of flexibility, most notably in the form of profiles in MPEG-2. However, they still code video, that is, a regular time sequence of two-dimensional video frames and audio samples. In contrast, MPEG-4 establishes a standard for the coding of two-dimensional and three-dimensional audiovisual objects. The MPEG-4 standard is being developed in two sequential versions. Version 1 is scheduled to be completed December 1998, while Version 2 is scheduled to be adopted one year later. Version 2 will be a subset of Version 1, thereby insuring

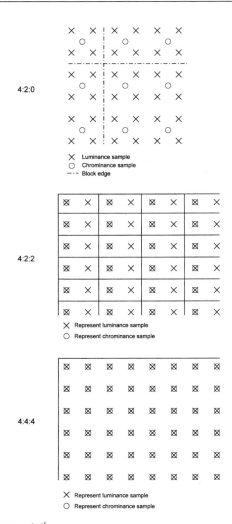

Figure 11.5 Color coding representations.

backward compatibility. For reference purposes, the official MPEG-4 standard designations are listed below.

14496-1   MPEG-4 Systems;

14496-2   MPEG-4 Visual Coding;

14496-3   MPEG-4 Audio Coding;

14496-4   MPEG-4 Conformance (to be standardized later);

14496-5   MPEG-4 Reference Software;

14496-6   MPEG-4 DMIF.

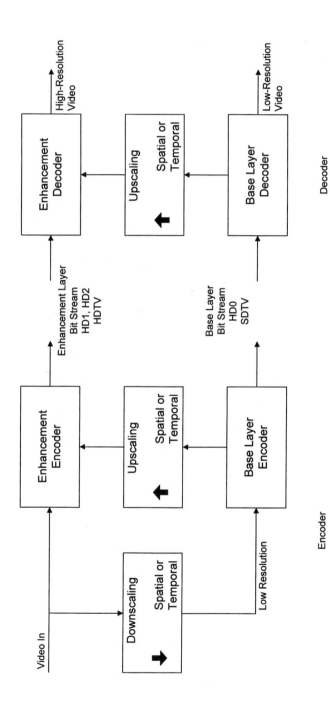

Figure 11.6 Scalable coding of video.

11.3.1   MPEG-4 Applications

Key applications for MPEG-4 are summarized below.

- *Real-time communication* systems are targeted toward applications that encompass two-way human interaction, or one-way applications that impose strict one-way delay constraints. A videophone system is a prime example of a two-way real-time system.

- *Surveillance:* Many modern surveillance sensors produce output in the form of images or video signals. Audio sensors are also used. In many applications, these sensors are connected via a telecommunications system to terminals that provide both monitoring and control.

- *Mobile multimedia* refers to the use of a portable computer capable of wireless communication. In a typical mobile computing scenario, a mobile user communicates with a remote computer system using a notebook or a personal digital assistant via wireless communications links.

- *Infotainment* refers to a combination of entertainment and information involving the interaction of the user with audiovisual objects (AVOs). Users would have the means to get information about a specific subject of interest (e.g., statistics on a sports player).

- *DVD* applications include interactive movies, travel, guidance, self-learning, games, and Internet Karaoke.

- *Content-based storage and retrieval* systems aim to provide access based on attributes associated with the video or audio content where these attributes may be keywords as well as numerical attributes.

- *Streaming video on the Internet/intranet* is an application which enables video transmission from a server to clients using the Internet. Different from a file transfer, video can be viewed immediately after receiving data without waiting for the entire file to download.

- *Broadcast* applications are defined by the presence of one or more broadband, unidirectional fixed-bandwidth channels providing multiple programs each of which may contain the full range of audio, video, and data content.

- *Digital television set-top box (DTV)* will change the nature of television broadcasting since digital data together with digital audio/video can be delivered to consumers. The interactive nature of MPEG-4 will make it a prime candidate for location in the set-top box to enable applications such as home-shopping, interactive game shows, and program guides.

- *Studio and television post-production* may be performed based on the content of the AV material.

- *Television modeling language (TVML)* is a script description language that can produce full TV programs in real time by using real-time computer graphic characters, a voice synthesizer, and multimedia computing techniques.

- *Collaborative scene visualization* supports a class of computer-supported cooperative work applications where groups of people who typically work simultaneously in distributed locations leverage visualization tools to accomplish a task by sharing a common visual information space.

- *Face animation application and virtual meeting:* People in a virtual meeting are each represented in a simple virtual environment by three-dimensional animated faces driven by video and/or audio.

- *Digital AM broadcasting* has the objective of improving the current broadcasting services within the AM bands by introducing digital programs in addition to the current analog programs.

11.3.2   MPEG-4 Systems Design

MPEG-4 provides standardized ways to:

1. Represent units of aural, visual, or audiovisual content, called "audiovisual objects" or AVOs. (The very basic unit is more precisely called a "primitive AVO.") These AVOs can be of natural or synthetic origin; this means they could be recorded with a camera or microphone, or generated with a computer.

2. Compose these objects together to create compound audiovisual objects that form audiovisual scenes.

3. Multiplex and synchronize the data associated with AVOs, so that they can be transported over network channels providing a QoS appropriate for the nature of the specific AVOs.

4. Interact with the audiovisual scene generated at the receiver's end.

Audiovisual scenes, such as that illustrated in Figure 11.7, are composed of several AVOs, organized in a hierarchical fashion. At the leaves of the hierarchy, we find primitive AVOs, such as:

- Still images (e.g., a fixed background);

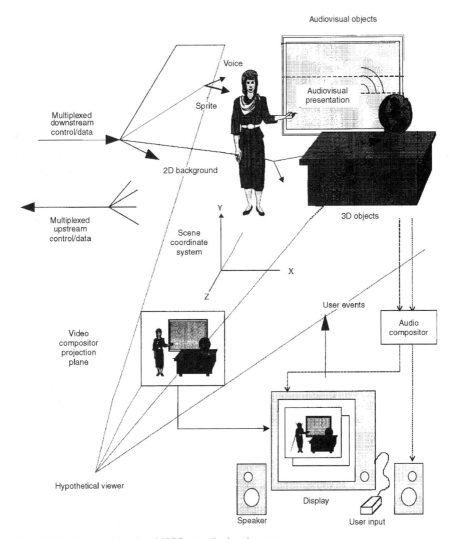

Figure 11.7 An example of an MPEG-4 audiovisual scene.

- Video objects (e.g., a talking person without the background);
- Audio objects (e.g., the voice associated with that person).

MPEG-4 standardizes a number of such primitive AVOs, capable of representing both natural and synthetic content types, which can be either two- or three-dimensional. In addition to the AVOs mentioned above and shown in Figure 11.7, MPEG-4 defines the coded representation of objects such as:

- Text and graphics;
- Talking synthetic heads and associated text to be used to synthesize the speech and animate the head;
- Synthetic sound.

Figure 11.7 gives an example that highlights the way in which an audiovisual scene in MPEG-4 is composed of individual objects. The figure contains compound AVOs that group elementary AVOs together. As an example: the visual object corresponding to the talking person and the corresponding voice are tied together to form a new compound AVO, containing both the aural and visual components of a talking person.

Such grouping allows authors to construct complex scenes, and enables consumers to manipulate meaningful (sets of) objects.

More generally, MPEG-4 provides a standardized way to compose a scene, allowing authors, for example, to:

- Place media objects (AVOs) anywhere in a given coordinate system;
- Apply transforms to change the geometrical or acoustical appearance of a media object;
- Group primitive media objects in order to form compound media objects;
- Apply streamed data to media objects, in order to modify their attributes (e.g., moving texture belonging to an object, animating a moving head by sending animation parameters);
- Change, interactively, the user's viewing and listening points anywhere in the scene.

The scene composition borrows several concepts from virtual reality modeling language (VRML) in terms of both its structure and the functionality of object composition nodes.

11.3.2.1 Description and Synchronization of Streaming Data for Media Objects

Media objects may rely on streaming data that is conveyed in one or more elementary streams. All streams associated to one media object are identified by an object descriptor. This allows for the handling of hierarchically encoded data as well as the association of meta information about the content (object content information) and the intellectual property rights associated with it. Each stream characterizes itself by a set of descriptors conveying configuration information (e.g., to determine the required decoder resources and the precision of encoded timing information). Furthermore, the descriptors may convey hints to the QoS it requests for transmission (e.g., maximum bit rate, bit error rate, priority). Synchronization of elementary streams is achieved through time stamping of individual access units within elementary streams. The identification of such access units and the time stamping is accomplished by the synchronization layer. Independent of the media type, this layer allows identification of access units (e.g., video or audio frames, scene description commands) in elementary streams and recovery of the media object's or scene description's time base, and enables synchronization among them. The syntax of this layer is configurable in a large number of ways, allowing use in a broad spectrum of systems.

11.3.2.2 Delivery of Streaming Data

The synchronized delivery of streaming information from source to destination, exploiting different QoS as available from the network, is specified in terms of the aforementioned synchronization layer and a delivery layer containing a two-layer multiplexer, as depicted in Figure 11.8.

The first multiplexing layer is managed according to the DMIF specification, part 6 of the MPEG-4 standard. This multiplex may be embodied by the MPEG-defined FlexMux tool, which allows grouping of elementary streams (ESs) with a low-multiplexing overhead. Multiplexing at this layer may be used, for example, to group ESs with similar QoS requirements or to reduce the number of network connections or the end to end delay.

The "TransMux" (transport multiplexing) layer in Figure 11.8 models the layer that offers transport services matching the requested QoS. Only the interface to this layer is specified by MPEG-4 while the concrete mapping of the data packets and control signaling must be done in collaboration with the bodies that have jurisdiction over the respective transport protocol. Any suitable existing transport protocol stack such as (RTP)/UDP/IP, (AAL5)/ATM, or MPEG-2's transport stream over a suitable link layer may become a specific TransMux instance. The choice is left to the end user/service provider, and allows MPEG-4 to be used in a wide variety of operation environments.

Use of the FlexMux multiplexing tool is optional and, as shown in Figure 11.8, this layer may be empty if the underlying TransMux instance provides all the required functionality. The synchronization layer, however, is always present. With regard to Figure 11.8, it will be possible to:

1. Identify access units, transport timestamps, and clock reference information and identify data loss;

2. Optionally interleave data from different elementary streams into FlexMux streams;

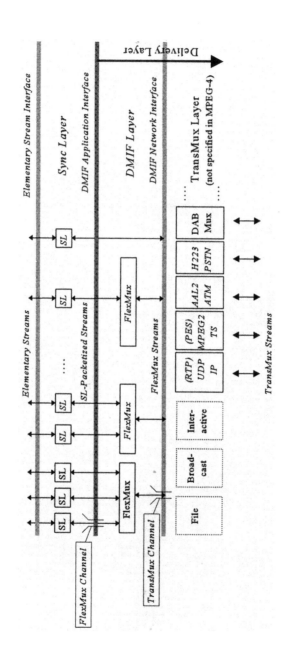

Figure 11.8 The MPEG-4 system layer model.

3. Convey control information to:

- Indicate the required QoS for each elementary stream and FlexMux stream;
- Translate such QoS requirements into actual network resources;
- Associate elementary streams to media objects;
- Convey the mapping of elementary streams to FlexMux and Trans-Mux channels.

Part of the control functionalities will be available only in conjunction with a transport control entity like the delivery multimedia integration framework (DMIF).

11.3.2.3 Interaction With Media Objects

In general, the user observes a scene that is designed by the scene's author. Depending on the degree of freedom allowed by the author, however, the user has the possibility to interact with the scene. Operations a user may be allowed to perform include:

- Change the viewing/listening point of the scene (e.g., by navigation through a scene);
- Drag objects in the scene to a different position;
- Trigger a cascade of events by clicking on a specific object (e.g., starting or stopping a videostream);
- Select the desired language when multiple language tracks are available;
- More complex kinds of behavior can also be triggered (e.g., a virtual phone rings, the user answers and a communication link is established).

As shown in Figure 11.9, streams coming from the network (or a storage device) as TransMux streams are demultiplexed into FlexMux streams and passed to appropriate FlexMux demultiplexers that retrieve ESs. The ESs are parsed and passed to the appropriate decoders. Decoding recovers the data in an AVO from their encoded form and performs the necessary operations to reconstruct the original AVO ready for rendering on the appropriate device. The reconstructed AVO is made available to the composition layer for potential use during scene rendering. Decoded AVOs, along with scene description information, are used to compose the scene as described by the author. The user can, to the extent allowed by the author, interact with the scene which is eventually rendered and presented.

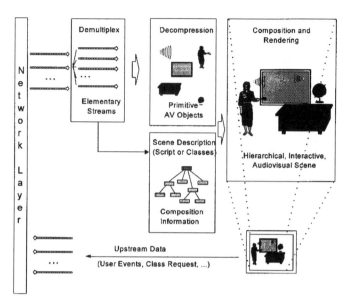

Figure 11.9 Major components of an MPEG-4 terminal (receiver side).

11.3.2.4 Scene Description

In addition to providing support for coding individual objects, MPEG-4 also provides facilities to compose a set of such objects into a scene. The necessary composition information forms the scene description, which is coded and transmitted together with the AVOs.

In order to facilitate the development of authoring, manipulation, and interaction tools, scene descriptions are coded independently from streams related to primitive media objects. Special care is devoted to the identification of the parameters belonging to the scene description. This is done by differentiating parameters that are used to improve the coding efficiency of an object (e.g., motion vectors in video-coding algorithms), and the ones that are used as modifiers of an object (e.g., the position of the object in the scene). Since MPEG-4 should allow the modification of this latter set of parameters without having to decode the primitive media objects themselves, these parameters are placed in the scene description and not in primitive media objects.

The following list gives some examples of the information described in a scene description.

- *How objects are grouped together:* An MPEG-4 scene follows a hierarchical structure that can be represented as a directed acyclic graph.

Each node of the graph is an AVO, as illustrated in Figure 11.10 (note that this tree refers back to Figure 11.7). The tree structure is not necessarily static; node attributes (e.g., positioning parameters) can be changed while nodes can be added, replaced, or removed.

- *How objects are positioned in space and time:* In the MPEG-4 model, AVOs have both a spatial and a temporal extent. Each AVO has a local coordinate system. A local coordinate system for an object is one in which the object has a fixed spatio-temporal location and scale. The local coordinate system serves as a handle for manipulating the AVO in space and time. AVOs are positioned in a scene by specifying a coordinate transformation from the object's local coordinate system into a global coordinate system defined by one more parent scene description nodes in the tree.

- *Attribute value selection:* Individual AVOs and scene description nodes expose a set of parameters to the composition layer through which part of their behavior can be controlled. Examples include the pitch of a sound, the color for a synthetic object, and activation or deactivation of enhancement information for scalable coding.

- *Other transforms on AVOs:* The scene description structure and node semantics are heavily influenced by VRML, including its event model.

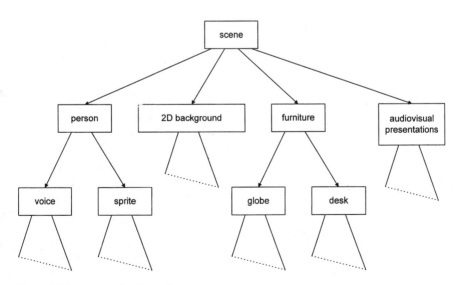

Figure 11.10 Logical structure of a scene.

This provides MPEG-4 with a very rich set of scene construction operators, including graphics primitives, that can be used to construct sophisticated scenes.

- *User interaction:* MPEG-4 allows for user interaction with the presented content. This interaction can be separated into two major categories: client-side interaction and server-side interaction. Client-side interaction involves content manipulation, which is handled locally at the end-user's terminal and can take several forms. In particular, the modification of an attribute of a scene description node (e.g., changing the position of an object, making it visible or invisible, changing the font size of a synthetic text node), can be implemented by translating user events (e.g., mouse clicks or keyboard commands). Server-side interaction involves content manipulations that occur at the transmitting end, initiated by a user action. This, of course, requires the use of a back channel.

11.3.2.5   Delivery Multimedia Integration Framework (DMIF)

DMIF is a session protocol for the management of multimedia streaming over generic delivery technologies. In principle it is similar to file transfer protocol (FTP). The only (but essential) difference is that FTP returns data; DMIF returns pointers to where to get (streamed) data.

When FTP is run, the first action it performs is the setup of a session with the remote side. Later, files are selected and FTP sends a request to download them. The FTP peer then returns the files in a separate connection.

Similarly, when DMIF is run, the first action it performs is the setup of a session with the remote side. Later, streams are selected and DMIF sends a request to stream them. The DMIF peer then returns the pointers to the connections where the streams will be streamed (and establishes the connection themselves).

Compared to FTP, DMIF is both a framework and a protocol. The functionality provided by DMIF is expressed by an interface called DAI (DMIF-application interface) and translated into protocol messages. These protocol messages may differ based on the network on which they operate.

The QoS is also considered in the DMIF design, and the DAI allows the DMIF user to specify the requirements for the desired stream. It is then up to the DMIF implementation to make sure that the requirements are fulfilled. The DMIF specification provides hints on how to perform such tasks on a few network types (including the Internet).

The DAI is also used for accessing broadcast material and local files. This means that a single, uniform interface is defined to access multimedia contents on a multitude of delivery technologies.

As a consequence, it is appropriate to state that the integration framework of DMIF covers three major technologies: interactive network technology, broadcast technology, and disk technology.

The DMIF architecture is such that applications which rely on DMIF for communication do not have to be concerned with the underlying communication method. The implementation of DMIF takes care of the delivery technology details, presenting a simple interface to the application.

Figure 11.11 represents the above DMIF concept. An application accesses data through the DAI, irrespectively whether such data comes from a broadcast source, from local storage, or from a remote server. In all scenarios the local application only interacts through a uniform interface (DAI). Different DMIF instances will then translate the local application requests into specific messages to be delivered to the remote application, taking care of the peculiarities of the involved delivery technology. Similarly, data entering the terminal (from remote servers, broadcast networks, or local files) is uniformly delivered to the local application through the DAI.

11.3.3   Coding of Visual Objects

Visual objects can be either of natural or of synthetic origin.

11.3.3.1   Natural Textures, Images, and Video

The tools for representing natural video in the MPEG-4 visual standard aim at providing standardized core technologies allowing efficient storage, transmission, and manipulation of textures, images, and video data for multimedia

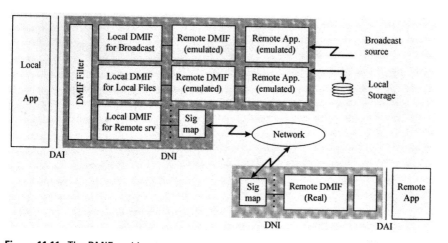

Figure 11.11 The DMIF architecture.

environments. These tools will allow the decoding and representation of atomic units of image and video content, called "video objects" (VOs). An example of a VO is a talking person (without background) which can then be composed with other AVOs to create a scene. Conventional rectangular imagery is handled as a special case of such objects.

In order to achieve this broad goal rather than a solution for a narrow set of applications, functionalities common to several applications are clustered. Therefore, the visual part of the MPEG-4 standard provides solutions in the form of tools and algorithms for:

- Efficient compression of images and video;
- Efficient compression of textures for texture mapping on two-dimensional and three-dimensional meshes;
- Efficient compression of implicit two-dimensional meshes;
- Efficient compression of time-varying geometry streams that animate meshes;
- Efficient random access to all types of visual objects;
- Extended manipulation functionality for images and video sequences;
- Content-based coding of images and video;
- Content-based scalability of textures, images, and video;
- Spatial, temporal, and quality scalability;
- Error robustness and resilience in error-prone environments.

The visual part of the MPEG-4 standard will provide a toolbox containing tools and algorithms bringing solutions to the above mentioned functionalities and more.

11.3.3.2 Synthetic Objects

Synthetic objects form a subset of the larger class of computer graphics. As an initial focus, the following visual synthetic objects will be described:

1. Parametric descriptions of:

   - A synthetic description of human face and body;
   - Animation streams of the face and body;

2. Static and dynamic mesh coding with texture mapping;

   - Texture coding for view dependent applications.

11.3.3.3 Structure of the Tools for Representing Natural Video

The MPEG-4 image and video coding algorithms will give an efficient representation of visual objects of arbitrary shape, with the goal to support so-called content-based functionalities. Next to this, it will support most functionalities already provided by MPEG-1 and MPEG-2, including the provision to efficiently compress standard rectangular-sized image sequences at varying levels of input formats, frame rates, pixel depth, bit rates, and various levels of spatial, temporal, and quality scalability.

A basic classification of the bit rates and functionalities currently provided by the MPEG-4 visual standard for natural images and video is depicted in Figure 11.12, with the attempt to cluster bit-rate levels versus sets of functionalities.

At the bottom end, a "VLBV core" (VLBV: very low bit-rate video) provides algorithms and tools for applications operating at bit rates typically between 5 to 64 Kbps and supporting image sequences with low spatial resolution (typically up to CIF resolution) and low frame rates (typically up to 15 Hz). The basic applications and specific functionalities supported by the VLBV core include:

- VLBV coding of conventional rectangular-size image sequences with high coding efficiency and high error robustness/resilience, low latency, and low complexity for real-time multimedia communication applications;
- Provisions for "random access" and "fast forward" and "fast reverse" operations for VLB multimedia data-base storage and access applications.

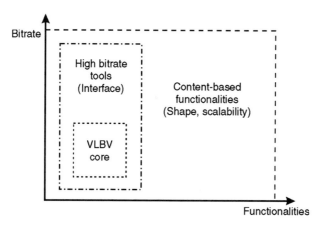

Figure 11.12 Classification of the MPEG-4 image and video coding algorithms and tools.

The same basic functionalities outlined above are also supported at higher bit rates with a higher range of spatial and temporal input parameters up to ITU-R Recommendation 601, resolutions employing identical or similar algorithms and tools as the VLBV core. The bit rates envisioned range typically from 64 Kbps up to 4 Mbps, and applications envisioned include broadcast or interactive retrieval of signals with a quality comparable to digital TV. For these applications at higher bit rates, tools for coding interlaced signals are specified in MPEG-4.

Content-based functionalities support the separate encoding and decoding of content (i.e., physical objects in a scene, VOs). This MPEG-4 feature provides the most elementary mechanism for interactivity—flexible representation and manipulation with or of VO content of images or video in the compressed domain, without the need for further segmentation or transcoding at the receiver.

For the hybrid coding of natural as well as synthetic visual data (e.g., for virtual presence or virtual environments), the content-based coding functionality allows mixing a number of VOs from different sources with synthetic objects, such as a virtual background.

The extended MPEG-4 algorithms and tools for content-based functionalities can be seen as a superset of the VLBV core and high bit-rate tools, meaning that the tools provided by the VLBV and high bit-rate video (HBV) cores are complemented by additional elements.

11.3.3.4   Support for Conventional and Content-Based Functionalities

The MPEG-4 video standard will support the decoding of conventional rectangular images and video as well as the decoding of images and video of arbitrary shape.

The coding of conventional images and video is achieved similar to conventional MPEG-1/2 coding and involves motion prediction/compensation followed by texture coding. For the content-based functionalities, where the image sequence input may be of arbitrary shape and location, this approach is extended by also coding shape and transparency information. Shape may be represented either by an 8-bit transparency component—which allows the description of transparency if one VO is composed with other objects—or by a binary mask.

The extended MPEG-4 content-based approach can be seen as a logical extension of the conventional MPEG-4 VLBV core or high bit-rate tools towards input of arbitrary shape.

11.3.3.5   The MPEG-4 Video Image and Coding Scheme

Figure 11.13 outlines the basic approach of the MPEG-4 video algorithms to encode rectangular as well as arbitrarily shaped input image sequences.

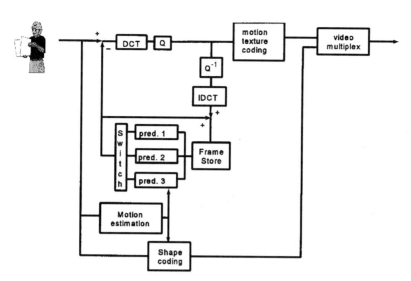

Figure 11.13 Basic block diagram of MPEG-4 video coder.

The basic coding structure involves shape coding (for arbitrarily shaped VOs) and motion compensation as well as DCT-based texture coding (using standard 8 × 8 DCT or shape adaptive DCT).

An important advantage of the content-based coding approach taken by MPEG-4 is that the compression efficiency can be significantly improved for some video sequences by using appropriate and dedicated object-based motion prediction "tools" for each object in a scene. A number of motion prediction techniques can be used to allow efficient coding and flexible presentation of the objects:

- Standard 8 × 8 or 16 × 16 pixel block-based motion estimation and compensation.

- Global motion compensation using eight motion parameters that describe an affine transformation.

- Global motion compensation based on the transmission of a static "sprite." A static sprite is a possibly large still image, describing panoramic background. For each consecutive image in a sequence, only eight global motion parameters describing camera motion are coded to reconstruct the object. These parameters represent the appropriate affine transform of the sprite transmitted in the first frame.

- Global motion compensation based on dynamic sprites. Sprites are not transmitted with the first frame but dynamically generated over the scene.

The structure of the visual coder is organized according to object profiles and combination profiles as shown in Tables 11.2 and 11.3, respectively.

11.3.4   MPEG-4 Audio Coding

MPEG-4 coding of audio objects provides tools both for representing natural sounds (such as speech and music) and for synthesizing sounds based on structured descriptions. The representation for synthesized sound can be derived from text data or so-called instrument descriptions, and by coding parameters to provide effects, such as reverberation and spatialization. The representations provide compression and other functionalities, such as scalability or play-back at different speeds.

11.3.4.1   Natural Sound

MPEG-4 standardizes natural audio coding at bit rates ranging from 2 Kbps up to and above 64 Kbps. The presence of the MPEG-2 AAC standard within the MPEG-4 tool set will provide for general compression of audio in the upper bit-rate range. For these, the MPEG-4 standard defines the bitstream syntax and decoding processes in terms of a set of tools. In order to achieve the highest audio quality within the full range of bit rates and at the same time provide the extra functionalities, three types of coding structures have been incorporated into the standard.

1. Parametric coding techniques cover the lowest bit-rate range (i.e., 2–4 Kbps for speech with 8-kHz sampling frequency and 4–16 Kbps for audio with 8- or 16-kHz sampling frequency).

2. Speech coding at medium bit rates between about 6–24 Kbps uses CELP coding techniques. In this region, two sampling rates, 8 and 16 kHz, are used to support both narrowband and wideband speech, respectively.

3. For bit rates starting below 16 Kbps, time-to-frequency (T/F) coding techniques, namely the TwinVQ and AAC codecs, are applied. The audio signals in this region typically have sampling frequencies starting at 8 kHz.

To allow optimum coverage of the bit rates and to allow for bit-rate and bandwidth scalability, a general framework has been defined. This is illustrated in Figure 11.14.

Starting with a coder operating at a low bit rate, by adding enhancements, such as the addition of bit-sliced arithmetic coding (BSAC) to an AAC coder for fine-grain scalability, both the coding quality and the audio bandwidth can

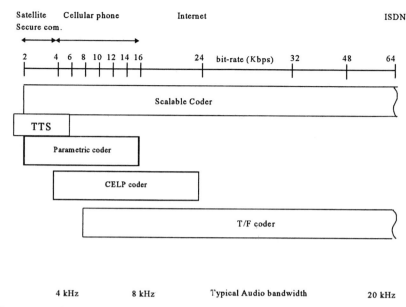

Figure 11.14 General block diagram of MPEG-4 audio.

be improved. These enhancements are realized within a single coder or, alternatively, by combining different coding techniques.

Bit-rate scalability allows a bitstream to be parsed into a bitstream of lower bit rate that can still be decoded into a meaningful signal. The bitstream parsing can occur either during transmission or in the decoder. Bandwidth scalability is a particular case of bit-rate scalability whereby part of a bitstream representing a part of the frequency spectrum can be discarded during transmission or decoding. Encoder complexity scalability allows encoders of different complexity to generate valid and meaningful bitstreams.

The decoder-complexity scalability allows a given bitstream to be decoded by decoders of different levels of complexity. The audio quality, in general, is related to the complexity of the encoder and decoder used. Error robustness provides the ability for a decoder to avoid or conceal audible distortion caused by transmission errors.

11.3.4.2 Synthesized Sound

Decoders are also available for generating sound based on structured inputs. Text input is converted to speech in the text-to-speech (TTS) decoder, while more general sounds including music may be normatively synthesized. Synthetic music may be delivered at extremely low bit rates while still describing an exact sound signal.

Table 11.2
Visual Object Profiles

Visual Tools	Visual Object Profiles									
	Simple	Core	Main	Simple Scale	12-Bit	Basic Anim. 2D Texture	Anim. 2D Mesh	Simple Face	Simple Scalable Texture	Core Scalable Texture
Intra Coding Mode (I-VOP)	X	X	X	X	X		X			
Inter Prediction Mode (P-VOP)	X	X	X	X	X		X			
AC/DC Prediction	X	X	X	X	X		X			
Slice Resynchronization	X	X	X	X	X		X			
Data Partitioning	X	X	X	X	X		X			
Reversible VLC	X	X	X	X	X		X			
4MV, Unrestricted MV	X	X	X		X		X			
Binary Shape Coding		X	X		X	X	X			
H.263/MPEG2 Quantization Tables		X	X	X	X		X			
P-VOP Based Temporal Scalability Rectangular Shape (JNB)		X	X	X	X		X			
P-VOP Based Temporal Scalability Arbitrary Shape (JNB)		X	X		X					
Bi-directional Pred. Mode (B-VOP)		X	X	X			X			
OBMC			X							

Table 11.2 *(continued)*

	1	2	3	4	5	6	7	8
Temporal Scalability Rectangular Shape		X						
Temporal Scalability Arbitrary Shape								
Spatial Scalability Rectangular Shape		X						
Static Sprites (Includes Low Latency Mode)	X							
Interlaced Tools	X							
Grayscale Alpha Shape Coding	X							
4- to 12-Bit Pixel Depth			X					
2D Dynamic Mesh With Uniform Topology				X	X			
2D Dynamic Mesh With Delaunay Topology					X			
Facial Animation Parameters						X		
Facial Interpolation Table (in Systems)								
Facial Definition Parameters (in Systems)								
Scalable Wavelet Texture (Rectangular, Spatial, & SNR Scalable) (USNB)				X	X			X
Scalable Wavelet Texture (Spatial Scalable) (JNB)				X	X		X	X
Scalable Wavelet Texture (All Tools, Including Shape Adaptive) (JNB)				X	X			
Facial Animation Tables								

Table 11.3
Visual Combination Profiles

Combination Profiles		Simple	Core	Main	Simple Scale	12-Bit	Basic Anim. 2D Texture	Anim. 2D Mesh	Simple Face	Simple Scalable Texture	Core Scalable Texture
1	Simple	X									
2	Simple B-VOP Scalable	X			X						
3	Core	X	X								
4	Main	X	X	X						X	X
5	12-Bit	X	X			X					
6	Simple Scalable Texture									X	
7	Simple FA								X		
8	Hybrid	X	X				X	X	X	X	X
9	Basic Animated 2D Texture						X		X	X	X

TTS coders' bit rates range from 200 bps to 1.2 Kbps, allowing a text or a text with prosodic parameters (pitch contour, phoneme duration, and so on) as its inputs to generate intelligible synthetic speech. It includes the following functionalities:

- Speech synthesis using the prosody of the original speech;
- Lip synchronization control with phoneme information;
- Trick mode functionality: pause, resume, jump forward or backward;
- International language and dialect support for text (i.e., the language and dialect that should be used can be signaled in the bitstream);
- International symbol support for phonemes;
- Support for specifying age, gender, speech rate of the speaker.

MPEG-4 audio coding is structural according to the object and combination profiles illustrated in Tables 11.4 and 11.5, respectively.

11.3.5   Intellectual Property Management and Protection

The MPEG-4 standard specifies a multimedia bitstream syntax and a set of tools and interfaces for designers and builders of a wide variety of multimedia applications. Each of these applications has a set of requirements regarding protection of the information it manages. These applications can produce conflicting content management and protection requirements. For some applications, users exchange information that has no intrinsic value but that must still be protected to preserve various rights of privacy. For other applications, the managed information has great value to its creator and/or distributors, requiring high-grade management and protection mechanisms. The implication is that the design of the intellectual property management and protection (IPMP) framework needs to consider the complexity of the MPEG-4 standard and the diversity of its applications.

The intellectual property management and protection methods required are as diverse as these applications. That is, the level and type of protection required depends on the content's value, complexity, and the sophistication of the associated business models. It is also possible that inter-domain security conflicts will arise, but their control should not be exercised through the standard but rather through other mechanisms associated with each application. The MPEG-4 IPMP framework is designed with these types of conflicts and needs in mind, providing application builders with the ability to construct the most appropriate domain-specific IPMP solution.

Table 11.4
Audio Object Profiles

Profile	Hierarchy	Tools Supported
AAC Main	Contains AAC LC	13818-7 (AAC) main profile PNS
AAC LC		13818-7 Low Complexity profile PNS
AAC SSR		13818-7 SSR profile PNS
T/F		13818-7 LC PNS LTP
T/F Main scalable	Contains T/F LC scalable	13818-7 main PNS LTP BSAC tools for large step scalability (TLSS) core codecs: CELP, TwinVQ, HILN
T/F LC scalable		13818-7 LC PNS LTP BASAC tools for large step scalability (TLSS) core codecs: CELP, TwinVQ, HILN
TwinVQ core		TwinVQ
CELP		CELP
HVXC		HVXC
HILN		HILN
TTSI		Text-To-Speech Interface
Main Synthetic	Contains Wavetable Synth.	All structured audio tools
Wavetable Synthesis		SASBF MIDI

Figure 11.15 illustrates the modular IPMP approach. It shows a clear point of separation between non-normative IPMP systems and the normative part of MPEG-4. This point of separation is the IPMP interface.

This approach allows the design of domain-specific IPMP systems (IPMP-Ss). While MPEG-4 does not standardize IPMP systems, it does standardize the MPEG-4 IPMP interface. This interface consists of IPMP descriptors (IPMP-Ds) and IPMP elementary streams (IPMP-ESs).

Table 11.5
Audio Combination Profiles

Combination Profile	Hierarchy	Audio Object Profiles Supported
Main	Contains Scalable, Speech and Low Rate Synthetic	AAC Main, LC, SSR T/F, T/F Main Scalable, T/F LC Scalable TwinVQ core CELP HVXC HILN Main Synthetic TTSI
Scalable	Contains Speech	T/F LC Scalable AAC-LC or/and T/F CELP HVXC TwinVQ core HILN Wavetable Synthesis TTSI
Speech		CELP HVXC TTSI
Low Rate Synthesis		Wavetable Synthesis TTSI

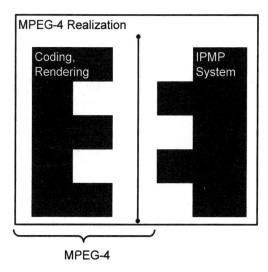

Figure 11.15 High level view of the proposed IPMP architecture.

IPMP-Ds and IPMP-ESs provide a communication mechanism between IPMP systems and the MPEG-4 terminal. Certain applications may require multiple IPMP-Ss. When MPEG-4 objects require management and protection, they have IPMP-Ds associated with them. These IPMP-Ds indicate which IPMP-Ss are to be used and provide information to these systems about how to manage and protect the content.

Figure 11.16 shows that the IPMP interface follows closely the MPEG-4 object/stream model. It has been designed to allow applications designers maximal flexibility, including the option of bypassing the mechanism altogether.

11.4   MPEG-7

The purpose of the MPEG-7 project (entitled "Multimedia Content Description Interface") is to develop a solution to the problem of quickly and efficiently searching for various types of multimedia material. MPEG-7 will specify a standard set of descriptors that can be used to describe various types of multimedia information. This description (i.e., the combination of descriptors and description schemes) is associated with the content itself, to allow fast and efficient searching for material of a user's interest. This "material" can include: still pictures, graphics, 3D models, audio, speech, and video. Descriptors can have a low abstraction level (e.g., shape, size, texture, color, trajectory) or they may

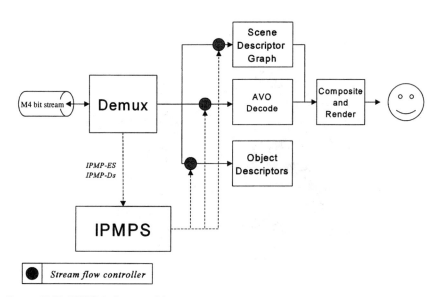

Figure 11.16 MPEG-4 player architecture.

have a high-level semantic form (e.g., "This is a scene with a barking brown dog on the left and a blue ball on the right with the sound of thunder in the background"). Low-level features can be extracted in fully automatic ways, whereas high-level features require more human interaction. MPEG-7 data can be physically located with the associated audiovisual material or located elsewhere. Neither the feature extraction process, nor the search engine is part of the MPEG-7 standard as illustrated in Figure 11.17.

MPEG-7 began its life as a scheme for making audiovisual material "as searchable as text is today." Although the proposed multimedia content descriptions are now acknowledged to serve much more than search applications, they remain for many the primary application for MPEG-7. These retrieval, or "pull" applications, involve databases, audiovisual archives, and the web-based Internet paradigm (a client requests material from a server).

In contrast with the above "pull" applications, the following "push" applications follow a paradigm more akin to broadcasting and the emerging web-casting. The paradigm moves from indexing and retrieval, as above, to selection and filtering. Such applications have very distinct requirements, generally dealing with streamed descriptions rather than static descriptions stored on databases.

A new abstract representation of possible applications using MPEG-7 is illustrated in Figure 11.18. The tentative time schedule for MPEG-7 is provided in Table 11.6.

11.5 JPEG Coding Algorithm

The JPEG is an ISO/ITU working group that developed an international standard ("Digital Compression and Coding of Continuous-Tone Still Images") for general purpose, continuous-tone (grayscale or color), still image compression. The aim of the standard algorithm is to be general purpose in scope to support a wide variety of image communication services. JPEG reports jointly to both the ISO group responsible for Coded Representation of Picture and Audio Information (ISO/IEC JTC1/SC2/WG8) and to the ITU Special Rapporteur group for Common Components for Image Communication (a subgroup of ITU SGX). This dual reporting structure is intended to ensure that the ISO and the ITU reduce compatible image compression standards.

The JPEG draft standard specifies two classes of encoding and decoding processes, lossy and lossless processes. Those based on the DCT are lossy, thereby allowing substantial compression to be achieved while producing a reconstructed image with high-visual fidelity to the encoder's source image. The simplest DCT-based coding process is referred to as the baseline sequential

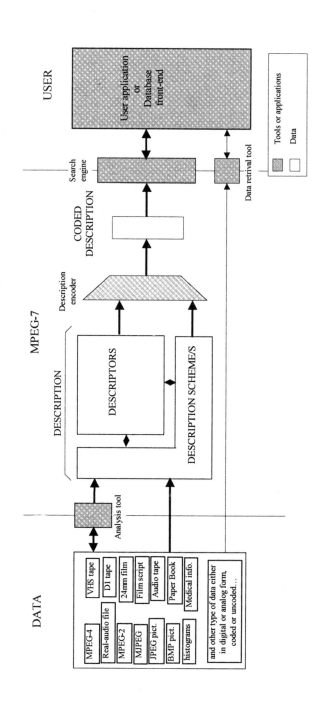

Figure 11.17 Interrelationship among defined concepts of MPEG-7.

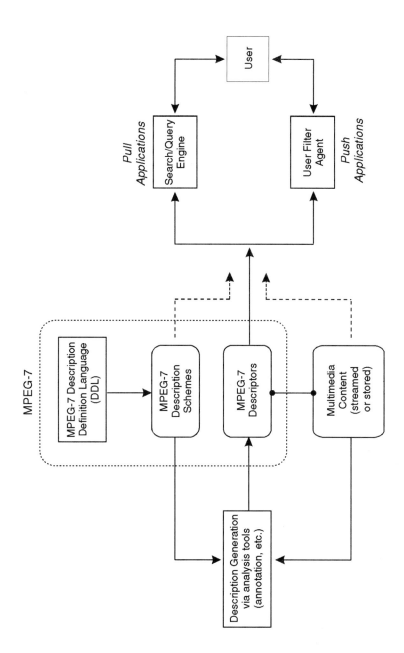

Figure 11.18 An abstract representation of possible applications using MPEG-7.

Table 11.6
MPEG-7 Time Schedule

October 16, 1998	**Call for Proposals** **Final Version of Proposal Package Desription**
February 1	Proposals due
February 15–19, 1999	Evaluation of proposals (in an Ad Hoc Group meeting)
March 1999	First version of MPEG-7 Verification Model
December 1999	Working Draft
October 2000	Committee Draft
February 2001	Final Committee Draft
July 2001	Final Draft International Standard
September 2001	International Standard

process. It provides a capability that is sufficient for many applications. There are additional DCT-based processes that extend the baseline sequential process to a broader range of applications. In any application environment using extended DCT-based decoding processes, the baseline decoding process must be present to provide a default decoding capability. The second class of coding processes is not based upon the DCT and is provided to meet the needs of applications requiring lossless compression (e.g., medical X-ray imagery). These lossless encoding and decoding processes are used independently of any of the DCT-based processes.

11.5.1   The Baseline System

Baseline system is the name given to the simplest image coding/decoding capability proposed for the JPEG standard. It consists of techniques that are well known to the image coding community, including 8 × 8 DCT, uniform quantization, and Huffman coding. Together these provide a lossy, high-compression image coding capability, which preserves good image fidelity at high compression rates.

The baseline system provides sequential build-up only. The baseline system codes an image to full quality in one pass and is geared toward line-by-line scanners, printers, and group 4 facsimile machines. Typically, the process starts at the top of the image and finishes at the bottom, allowing the recreated image to be built up on a line-by-line basis. One advantage is that only a small part of the image is being buffered at any given moment. Another feature stipulates that the recreated image need not be an exact copy of the original—the idea being that an almost indistinguishable copy of the original is just as good as an exact copy for most purposes. By not requiring exact copies, higher compression, which translates into lower transmission times, can be realized. Together these features are known as lossy sequential coding or transmission.

Figure 11.19 shows the main procedures for all encoding processes based on the DCT. It illustrates the special case of single component image (as opposed to multiple component color images); this is an appropriate simplification for overview purposes because all processes specified in this international standard operate on each image component independently.

In the encoding process, the input component's samples are grouped into 8 × 8 blocks, and each block is transformed by the forward DCT (FDCT) into a set of 64 values referred to as DCT coefficients. One of these values is referred to as the DC coefficient, and the other 63 as the AC coefficients.

Each of the 64 coefficients is then quantized using one of 64 corresponding values from a quantization table (determined by one of the table specifications shown in Figure 11.19). No default values for quantization tables are specified in this international standard; applications may specify values that customize picture quality for their particular image characteristics, display devices, and viewing conditions.

After quantization, the DC coefficient and the 63 AC coefficients are prepared for entropy encoding. The previous quantized DC coefficient is used to predict the current quantized DC coefficient, and the difference is encoded. The 63 quantized AC coefficients undergo no such differential encoding but are converted into a one-dimensional zigzag sequence that is common for DCT coding.

All of the quantized coefficients are then passed to an entropy encoding procedure, which compresses the data further. Since Huffman coding is used in the baseline system, Huffman table specifications must be provided to the encoder, as indicated in Figure 11.19.

Huffman coding has two forms, namely, fixed and adaptive. Fixed Huffman coding assumes that coding tables can be generated in advance from test images and then used for many images. In adaptive Huffman coding, the encoder analyzes an image's statistics before coding and devises Huffman tables tailored to that image. These tables are then transmitted to the decoder. Then the image is coded and transmitted. Upon receipt, the decoder can reconstruct the image using the previously transmitted, tailor-made, Huffman tables.

Figure 11.20 shows the main procedures for all DCT-based decoding processes. Each step shown performs essentially the inverse of its corresponding main procedure within the encoder. The entropy decoder decodes the zigzag sequence of quantized DCT coefficients. After dequantization, the DCT coefficients are transformed to an 8 × 8 block of samples by the inverse DCT

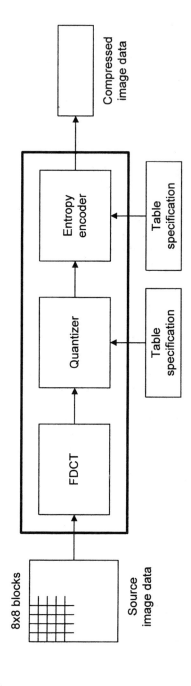

Figure 11.19 DCT-based encoder simplified diagram.

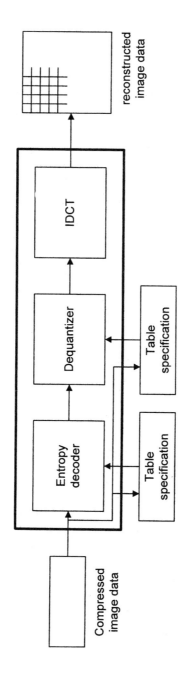

Figure 11.20 DCT-based decoder simplified diagram.

(IDCT). For DCT-based processes, two alternative sample precisions are specified, namely, either 8 or 12 bits per sample. The baseline process uses only 8-bit precision.

11.5.2 Extended System

Extended system is the name given to a set of additional capabilities not provided by the baseline system. Each set is intended to work in conjunction with, and to build upon, the components internal to the baseline system in order to extend its modes of operation. These optional capabilities, which include arithmetic coding, progressive build-up, progressive lossless coding, and others, may be implemented singly or in appropriate combinations.

Arithmetic coding is an optional, "modern" alternative to Huffman coding. Because the arithmetic coding method chosen adapts to image statistics as it encodes, it generally provides 5% to 10% better compression than the Huffman method chosen by JPEG. This benefit is balanced by some increase in complexity.

Progressive build-up, the alternative to sequential build-up, is especially useful for human interaction with picture databases over low-bandwidth channels. For progressive coding, a coarse image is sent first, then refinements are sent, improving the coarse image's quality until the desired quality is achieved. This process is geared toward applications such as image databases with multiple resolution and quality requirements, freeze-frame teleconferencing, photo-videotex over low-speed lines, and database browsing. There are three different, complementary, progressive extensions—spectral selection, successive approximations, and hierarchical.

Progressive lossless refers to a lossless compression method that operates in conjunction with progressive build-up. In this mode of operation, the final stage of progressive build-up results in a received image that is bit-for-bit identical to the original.

The JPEG draft standard includes the requirement that the baseline system be contained within every JPEG-standard codec, which utilizes any of the extended system capabilities. In this way, the baseline system can serve as a default communications mode for services that allow encoders and decoders to negotiate. In such cases, image communicability between any JPEG sender and receiver that are not equipped with a common set of extended system capabilities is assured. The operation of JPEG in the baseline and extended modes is summarized in Table 11.7. JPEG has been used for the transmission of grayscale pictures for many years, and a general consensus has developed regarding its coding efficiency as summarized in Table 11.8. It should be noted that performance can be improved by employing advanced nonstandard versions of JPEG.

Table 11.7
JPEG Operating Modes

	Baseline	**Extended**
Transmission mode	Sequential	Progressive
Reversibility	Usually lossy	Can be lossless
Max input/output coding precision per pixel	Up to 8 bits	Up to 16 bits
Variable length coding	Huffman	Arithmetic
Applications	G4 fax, Videotex, Freeze Frame	Remote access to data base

11.5.3 Lossless Coding

Figure 11.21 shows the main procedures for the lossless encoding processes. A predictor combines the values of up to three neighborhood samples (A, B, and C) to form a prediction of the sample indicated by X in Figure 11.22. This prediction is then subtracted from the actual value of sample X, and the difference is losslessly entropy-coded by either Huffman or arithmetic coding.

11.5.4 JPEG 2000

The ISO has established a new project, designated as JPEG 2000, to develop a new coding standard for still images including bilevel, grayscale, and color pictures. Objectives of the program include improved compression performance relative to the existing JPEG system, the inclusion of lossless and lossy algorithms, and robust performance in a noisy environment. Intended applications include low bandwidth image dissemination, medical imagery, client/server communication over the Internet, libraries, facsimile, and security. The wavelet coding

Table 11.8
JPEG Performance

Bits/pixel	**Quality**
≥ 2	Indistinguishable
1.5	Excellent
0.75	Very good
0.50	Good
0.25	Fair

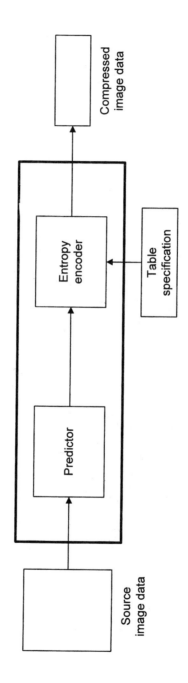

Figure 11.21 Lossless encoder simplified diagram.

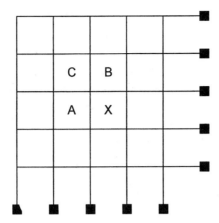

Figure 11.22 Three-sample neighborhood.

algorithm is a strong candidate for JPEG 2000. Core experiments comparing advanced DCT algorithms with wavelet coding have been completed, and it is likely that wavelets will play a major role in the new standard. The schedule calls for the completion of the Final Committee Draft and Draft International Standard in 1999 and full approval of the International Standard in the year 2000.

11.6 The JBIG Coding Algorithm

In 1988, an experts group was formed to establish an international standard for the coding of bilevel images. The Joint Bilevel Image Group (JBIG) is sponsored by the ISO (IEC/JTC1/SC2/WG9) and the ITU (SG 8). In 1993, the JBIG finalized the standard entitled "Progressive Bilevel Image Compression Standard," and much of the material in this section is derived from this document.

The JBIG standard defines a method for compressing a bilevel image (that is, an image that, like a black and white image, has only two colors). Because the method adapts to a wide range of image characteristics, it is a very robust coding technique. On scanned images of printed characters, observed compression ratios have been from 1.1 to 1.5 times as great as those achieved by the Modified READ (MMR) encoding of the ITU Recommendations T.4 (G3) and T.6 (G4). On computer generated images of printed characters, observed compression ratios have been as much as five times as great. On images with grayscale rendered by halftoning or dithering, observed compression ratios have been up to 30 times as great.

The method is bit-preserving, which means that it, like the ITU T.4 and T.6 recommendations, is distortionless and that the final decoded image is identical to the original.

The JBIG standard provides for both sequential and progressive operation. When decoding a progressively-coded image, a low-resolution rendition of the original image is made available first with subsequent doublings of resolution as more data is decoded. Progressive encoding has two distinct benefits. One is that one common database can efficiently serve output devices with widely different resolution capabilities. Only that information in the compressed image file that allows reconstruction to the resolution capability of the particular output device need be sent and decoded. Also, if additional resolution enhancement is desired for, say, a paper copy of something already on a CRT screen, only the needed updating information has to be sent.

The other benefit of progressive encoding is that it provides subjectively superior image browsing (on a CRT) over low-rate and medium-rate communication links. A low-resolution rendition is rapidly transmitted and displayed, with as much resolution enhancement as desired then following. Each stage of resolution enhancement builds on the image already available. Progressive encoding makes it easy for a user to quickly recognize the image being displayed, which in turn makes it possible for that user to quickly interrupt the transmission of an unwanted image.

Although the primary aim of this recommendation is bilevel image encoding, it is possible to effectively use this standard for multilevel image encoding by simply encoding each bit plane independently, as though it were itself a bilevel image. When the number of such bit planes is limited, such a scheme can be very efficient while at the same time providing for progressive buildup in both spatial refinement and grayscale refinement.

Acronyms and Abbreviations

The acronyms and abbreviations used in this document are as follows.

AAL	ATM adaptation layer
ACC	audio codec capabilities
ACELP	algebraic codebook excitation linear prediction
ACR	absolute category rating
ADPCM	adaptive differential code modulation
AL	adaption layer
ANSI	American National Standards Institute
ASCII	American Standard Code for Information Interchange
ATIS	Alliance for Telecommunications Industry Solutions
ATM	asynchronous transfer mode
AVO	audiovisual object
BAS	bit rate allocation signal
B-ISDN	broadband ISDN
BOF	Birds of a Feather
bps	bits per second
BRI	basic rate interface
BSAC	bit-sliced arithmetic coding
C&I	control and indication
CATS	Consortium for Audiographics Teleconferencing Standards
CCITT	International Telephone and Telegraph Consultative Committee
CCR	comparison category rating
CD	compact disc
CELP	codebook excited linear prediction
CIC	control and indication capabilities
CIF	common intermediate format
CODEC	coder-decoder
COMSEC	communications security crypto cryptographic

CPE	customer premises equipment
CS	convergence sublayer
CSU	channel service unit
CTS	clear to send
DAI	delivery multimedia integration framework application interface
DAT	digital audio tape
dBm	decibel referred to 1 milliwatt
DCE	data communication equipment
DCT	discrete cosine transform
DES	Data Encryption Standard
DFT	discrete Fourier transform
DHCP	dynamic host control protocol
DIS	Draft International Standard
DMIF	delivery multimedia integration framework
DPCM	differential pulse code modulation
DSI	digital speech interpolation
DSM-CC	digital storage media–command and control
DSP	digital signal processing
DSU	data service unit
DTE	data terminal equipment
DTV	digital television
DVC	desktop videoconferencing
EC	European Commission
ECSA	Exchange Carriers Standards Association
EIA	Electronic Industries Association
EG	engineering guideline
EMI	electromagnetic interference
EMP	electromagnetic pulse
EoB	end of block
ES	elementary stream
ESD	electrostatic discharge susceptibility
ETSI	European Telecommunications Standards Institute
FAP	facial animation parameter
FAS	frame alignment signal
FCIF	full common intermediate format
FDCT	forward DCT
FDDI	fiber distributed data interface
FDP	facial definition parameter
FEC	forward error correction
FTP	file transfer protocol
GATT	General Agreement on Tariffs and Trade

GCC	generic conference control
GIT	generic infornation transfer
GOB	group of blocks
GOP	group of pictures
GSTN	general switched telephone network
HDTV	high-definition TV
HSD	high-speed data
Hz	Hertz
I/O	input/output
IAB	Internet Architecture Board
IDCT	inverse DCT
IEC	International Electrotechnical Commission
IEEE	Institute of Electronic and Electrical Engineers
IESG	Internet Engineering Steering Group
IETF	Internet Engineering Task Force
IMTC	International Multimedia Teleconferencing Consortium
IPMP	intellectual property management and protection
IPMP-D	IPMP descriptors
IPMP-ES	IPMP elementary stream
IPMP-S	IPMP system
IRTF	Internet Research Task Force
IS	International Standard
ISDN	integrated services digital network
ISO	International Standards Organization
ISOC	Internet Society
ISONET	ISO Information Network
ITU	International Telecommunications Union
ITU-D	Telecommunication Development Sector (of the ITU)
ITU-R	Radiocommunications Sector (of the ITU)
ITU-T	Telecommunication Satandardization Sector (of the ITU)
IWU	interworking unit
JBIG	Joint Bilevel Image Group
JPEG	Joint Photographic Coding Experts Group
Kbps	kilobits per second
KLT	Karhunen-Loeve transfrom
LAN	local area network
LATA	local access and transport area
LCN	logical channel number
LD-CELP	low-delay code excited linear prediction
LPAS	linear prediction analysis-by-synthesis
LSD	low-speed data

kHz	kilohertz
Mbps	megabits per second
MC	multipoint controller; also multiplexes code
MCCOI	Multimedia Communications Community of Interest
MCNS	multipoint communication service
MCS	multipoint communications service
MCU	multipoint control unit
MILSTD	military standard
MIPS	million instructions per second
MLP	multilayer protocol
MOS	mean opinion score
MP	multipoint processor
MPEG	Motion Picture Experts Group
MPI	minimum picture interval
MPL	multiplex payload length
MP/ML	main profile/main level
MP-MLQ	MultiPulse-Maximum Likelihood Quantizer
ms	milliseconds
MUX	multiplex
N/A	not applicable
NAC	network adaptation capabilities
NAL	network adaptation layer
NATO	North Atlantic Treaty Organization
NBC	not backward compatible
N-ISDN	narrowband ISDN
NNI	network node interface
NTSC	National Television Standards Committee
OBASC	object-based analysis-synthesis coding
OBC	object-based coding
ODC	other data capabilities
PACS	Personal Advanced Communications Systems
PAL	phase alteration by line
PCM	pulse code modulation
PCWG	Personal Conferencing Work Group
PDU	protocol data unit
PES	packetized elementary stream
PID	packet ID
PPP	point-to-point protocol
PS	program stream
PSDN	packet-switched data network
PSTN	public-switched telephone network

QCIF	quarter common intermediate format
QDU	quantization distortion unit
QoS	quality of service
RAM	random access memory
RAS	Registration Admission Status
RAST	receive-and-send terminal
RE	receive end
RGB	red-green-blue
RISV	Reference Impairment System for Video
ROM	read only memory
ROT	receive-only terminal
RP	recommended practice
RSVP	resource reservation protocol
RTCP	real-time transport control protocol
RTP	real-time transport protocol
SAR	segmentation and reassembly
SB-ADPCM	subband adaptive differential pulse code modulation
SBM	subnet bandwidth management
SC	service channel
SCN	switched circuit network
SDH	synchronous digital hierarchy
SDT	structured data transfer
SECAM	sequential color avec memoir
SG	Study Group
SMPTE	Society of Motion Picture and Television Engineers
SOT	send-only terminal
SQEG	speech quality experts group
SRTS	synchronous residual time stamp
T1AG	T1 Advisory Group
TCP/IP	transmission control protocol/Internet protocol
TE	transmit end
T/F	time-to-frequency
TLS	transport layer security
TS	transport stream
TSAG	Telecommunication Standardization Advisory Group
TSB	Telecommunication Standardization Bureau
TSC	Technical Subcommittee
TTS	text-to-speech
TVML	television modeling language
UDP	user datagram protocol
VC	virtual channel

VCC	video codec capabilities
VLBV	very low bit-rate video
VLC	variable length code
VO	video object
VoIP	voice over IP
VP	videophone
VQ	vector quantization
VRML	virtual reality modeling language
VTC	video teleconferencing
WAN	wide area network
WD	Working Draft
WP	working party
WTSC	World Telecommunication Standardization Conference

Glossary

Adaptive differential pulse code modulation A form of differential pulse code modulation that uses adaptive quantizing.

A-law A PCM coding and companding standard used in Europe. One common use is for digital voice communications.

Analog-to-digital (A-D) conversion The conversion of an analog signal to a digital signal.

Asynchronous A transmission method in which units of data are sent one character at a time preceded and followed by start/stop bits that provide timing (synchronization) at the receive terminal.

Audiovisual Involving both hearing and sight.

Bandwidth The amount of information that can be sent over a given transmission channel, expressed typically as a frequency range or as bits per second. Roughly equated with speed. Typical transmission services are 64 and 128 Kbps, 1.544 Mbps (T1), and 45 Mbps (T3).

Basic rate Formally called the basic rate interface (BRI), one of two "interfaces" defined in ISDN. BRI provides two B-channels at 64 Kbps and a data D-channel at 16 Kbps. Defined for links to the home.

Baud Unit of digital transmission signaling speed derived from the duration of the shortest code element. Speed in baud is the number of code elements per second, also commonly referred to as symbols per second. Most modems today operate at 2400 baud. If the symbol captures 6 bits of information (V.32bis), then the 2400-baud modem can send 14,400 bits per second.

B-channel A "bearer" channel that is a fundamental component of ISDN and B-ISDN interfaces. It carries 64 Kbps in either direction, is circuit switched, and can carry either digitized voice or data bits.

Bit rate In a bitstream, the number of bits occurring per unit time, usually expressed as bits per second.

Bit rate allocation signal (BAS)   Bit position within the frame structure of ITU recommendation H.221 that is used to transmit commands, control and indication signals, and capabilities.

Bonding (Bandwidth-on-Demand Interoperability Group)   A group of electronic equipment manufacturers formed to promote the use of inverse multiplexers through the development of an operational standard. I-MUXes compliant with the same modes (or levels) of the bonding standard are now available from many manufacturers.

Bridge   A device for linking three or more channels of voice or data. Also, a computer that connects two or more networks and forwards packets among them. Usually, bridges operate at the physical network level.

Broadcast operation   The transmission of information so that it may be simultaneously received by stations that usually make no acknowledgment.

Camera   In television, an electronic device using an optical system and a light-sensitive pickup tube or chip to convert visual signals into electrical impulses.

Central office (CO)   Telephone company facility where subscriber lines are terminated on switching equipment, from which connections can be made to local and long distance points.

Channel   A single unidirectional or bidirectional path for transmitting or receiving, or both, of electrical or electromagnetic signals, usually in distinction from other parallel paths.

Channel service unit (CSU)   User-owned equipment installed on customer premises at the interface between customer premises and the operating phone company to terminate a circuit. CSUs provide network protection and diagnostic capabilities.

Chrominance   The difference between a reproduced color and a standard reference color of the same luminous intensity.

CIF   *See* Full common intermediate format (FCIF).

Circuit switching   A method of establishing a temporary, dedicated communications path between two or more locations through one or more switching nodes. Data are sent in a continuous stream; the data rate is constant; the delay is constant and limited to propagation times; and a dedicated end-to-end path remains in effect until the communication is terminated. Contrast with packet switching.

Classified Any information that has been determined to require protection against unauthorized disclosure to avoid harm to the United States national security. The classifications "top secret," "secret," and "confidential" are used to designate such information, referred to as classified information.

Codec Acronym for coder/decoder. An electronic device that converts analog signals, typically video, voice, and/or data, into digital form and compresses them into a fraction of their original size to save frequency bandwidth on a transmission path.

Communications A method or means of conveying information of any kind from one person or process to other person(s) or process(es) by a telecommunication medium.

Compression The application of any of several techniques that reduce the number of bits required to represent information in data transmission or storage, therefore conserving bandwidth and/or memory.

COMSEC equipment Equipment designed to provide security to telecommunications by converting information to a form unintelligible to an unauthorized interceptor and by reconverting such information to its original form for authorized recipients, as well as equipment designed specifically to aid in, or as an essential element of, the conversion process. Communications security equipment is cryptoequipment, cyrptoancillary equipment, cryptoproduction equipment, and authentication equipment.

Conferencing Programs and meetings that may be for the purpose of presenting and exchanging information, comparing views, learning, planning, and decision making. Conferences can be held in one location or conducted simultaneously at multiple locations and linked together by telecommunications systems. It includes the design and engineering of conferencing systems and telecommunications services; creation of presentation media; and the development and promulgation of policy, procedures, and standards for the operation of conferencing activities, facilities, systems, and networks.

Cryptography The principles, means, and methods for rendering plain information unintelligible and for restoring encrypted information to intelligible form.

Customer premises equipment (CPE) Terminal and associated equipment located at a subscriber's premise and connected with the termination of a carrier's communication channel(s) at the network interface at the subscriber's premises. Excluded from CPE are over-voltage protection equipment, inside wiring, coin-operated or pay telephones, and multiplexing equipment to deliver multiple channels to the customer.

Data   Representation of facts, concepts, or instructions in a formalized manner suitable for communication, interpretation, or processing by humans or by automatic means. A representation such as characters or analog quantities to which meaning is or might be assigned.

Data circuit-terminating equipment (DCE)   The interfacing equipment sometimes required to couple the data terminal equipment (DTE) into a transmission circuit or channel and from a transmission circuit or channel into the DTE.

Data communications port   A port used for the transfer of information between functional units by means of data transmission according to a protocol.

Data encryption standard (DES)   A national standard used in the United States for the encryption of information digitally transferred using a 64-bit key. The standard, which has been set by the National Bureau of Standards, provides only privacy protection and is not recognized by NSA as providing security protection.

Data rate   In digital data communications, the speed at which data (bits in this case) are transmitted, usually expressed in bits per second.

Data terminal equipment (DTE)   Equipment consisting of digital end instruments that convert the user information into data signals for transmission or reconvert the received data signals into user information.

Data transmission   Conveying data from one place for reception elsewhere by telecommunication means.

D-channel   Used to carry control signals and customer call data in ISDN. Basic rate interface provides a 16-Kbps D-channel, primary rate interface provides a 64-Kbps D-channel.

Decode   To convert data by reversing the effects of some previous encoding.

Decoder   A device that decodes. *See* decode.

Desktop and individual workstation   An input/output display device with local computer power allowing an individual to perform some computational work and database access from a local or remote location. It may also have videophone and/or VTC capabilities.

Differential pulse code modulation   A process in which a signal is sampled, and the difference between each sample of this signal and its estimated value is quantized and converted by encoding to a digital signal.

Discrete signals   A signal composed of sample values uniformly spaced in time. The result of sampling a continuous signal.

Dual tone multi-frequency (DTMF)   The series of two tone combinations generated by a telephone's keypad that is used for signaling within the system. Also known as a "touch tone" which is actually an AT&T service mark.

Duplex operation   An operating method in which transmission is permitted, simultaneously, in both directions of a telecommunication channel.

Echo   A wave that has been reflected or otherwise returned with sufficient magnitude and delay to be perceived.

Echo attenuation   In a communication circuit (4- or 2-wire) in which the two directions of transmission can be separated from each other, the attenuation of echo signals that return to the input of the circuit under consideration.

Echo cancellation   The process of reducing echo electronically in the audio system.

Echo canceller   A device that electronically reduces echo in the audio system.

Encoder   A device that encodes. *See* encoding.

Encoding   The process of reforming information into a format suitable for transmission.

Encrypt   To convert plain text into an unintelligible form by means of a cryptosystem.

Encryption   The process of encrypting.

Forward error correction (FEC)   A system of error control for data transmission wherein the receiving device has the capability to detect and correct any character or code block that contains fewer than a predetermined number of symbols in error. (FEC is accomplished by adding bits to each transmitted character or code block using a predetermined algorithm.)

Frame   The set of all the picture elements in an image.

Full common intermediate format (FCIF)   A video format defined in H.261 that is characterized by 352 luminance pixels on each of 288 lines, with half as many chrominance pixels in each direction.

Full duplex   An operating method in which transmission is permitted, simultaneously, in both directions of a telecommunication channel.

Gatekeeper (GK)   An H.323 entity that provides address translation and control access, and sometimes bandwidth management to the LAN for H.323 terminals, gateways, and MCUs.

Graphics   The art or science of conveying information through the use of, for example, graphs, letters, lines, drawings, and pictures.

Handshaking   The process used to establish communications parameters between two stations.

High-resolution graphics   A video system that provides better resolution than a standard home television. It is not unusual for a typewritten page to be easily read on a high-resolution monitor. Graphics with sufficient detail to be able to read a full 8 1/2- by 11-inch typed page; approximately 1,024 × 1,024 pixels of resolution.

Huffman coding   A general data compressor first described by D. A. Huffman in 1952. Huffman encoding is a method of encoding symbols that varies the length of the symbol in proportion to its information content.

Input/output (I/O) device (equipment)   A device that introduces data into or extracts data from a system.

Integrated services digital network (ISDN).   A project underway within the CCITT for the standardization of operating parameters and interfaces for a network that will allow a variety of mixed digital transmission services to be accommodated. Access channels include a basic rate (two 64-Kbps B channels + one 16-Kbps D channel) and a primary rate (23 64-Kbps B channels and one 64-Kbps D channel).

The International Telecommunications Union (ITU)   A new name for the CCITT. Founded in 1865, the organization is part of the UN and is composed of the telecommunications administrations of the participating nations. Focus is the maintenance and extension of international cooperation for improving telecommunications development and applications.

Internet protocol (IP)   The TCP/IP standard protocol that defines the IP datagram as the unit of information passed across an Internet and provides the basis for connectionless, best-effort packet delivery service. IP includes the ICMP control and error message protocol as an integral part. The entire protocol suite is often referred to as TCP/IP because TCP and IP are the two most fundamental protocols. IP is the network layer of the ISO seven-layer stack.

Interoperability   The condition achieved among communication-electronics systems or items of communications-electronics equipment when information or services can be exchanged directly and satisfactorily between them and/or to their users. The degree of interoperability should be defined when referring to specific cases.

JPEG   An industry standard compression method that was specifically designed for images. JPEG combines a quantizer, DCT, and either Huffman encoding or arithmetic coding to produce high compression ratios for color and gray scale images.

Key   A code that governs the encryption and decryption of information. Users must have the same key in order to decrypt each other's messages.

Leased line   A semi-permanent leased telephone circuit that connects two or more points and is continuously available to the subscriber.

Lip synchronization   The relative timing of audio and video signals so that there is no noticeable lag or lead between audio and video.

Local exchange carrier (LEC)   Local telephone company that provides subscriber lines and local calling services. Typically LECs have a monopoly on local service.

Local loop   A telephone circuit that connects a subscriber's station equipment to the switching equipment in the telephone company's central office. Also referred to as a subscriber loop.

Long-haul communications   Communication that permits users to convey information on a national or worldwide basis.

Luminance   The monochromatic signal used to convey brightness information.

Microphones   Devices that convert acoustic energy (sound waves) into electrical energy, to be transmitted over wire or other channels of communication. An audio transducer that converts sound pressure waves (sound energy) into electrical signals.

Modem (modulator/demodulator)   Device that connects terminals and hosts through analog links by converting digital signals to analog form and back again.

Monitors   *See* visual display unit.

Motion compensation coding   A type of interframe coding used by picture processors in the compression of video images. The process relies upon an algorithm that examines a sequence of frames to develop a prediction as to the motion that will occur in subsequent frames.

Mu-law   The PCM coding and companding standard used in Japan and North America.

Multiplexer   A device for combining two or more channels into a single channel.

Multipoint   A telecommunications system that allows each of three or more sites to both transmit and receive signals from all other sites.

Network   An interconnection of three or more communication entities and (usually) one or more nodes.

NTSC Standard   Acronym for National Television Standards Committee standard. North American standard for the generation, transmission, and reception of television communications wherein the 525-line picture is the standard versus the PAL and SECAM systems using the 625-line picture. (The picture information is transmitted in AM and the sound information is transmitted in FM. Compatible with CCIR Standard M.)

P × 64   Family of five ITU Recommendations. These include H.261, H.221, H.242, H.320, and H.230.

Packet   A bundle of data packaged for transmission over a network. Packets can be various lengths, ranging from about 40 bytes to 32,000 bytes on the Internet, but typically are about 1,500 bytes in length. ATM specifies 53 byte packets, known as cells, because the length is fixed. A packet is always associated with an address header and control information.

Parity   In binary-coded systems, the oddness or evenness of the number of ones in a finite binary system.

Pixel   A picture element that contains grayscale or color information. (Grayscale is an integration of density and gives resolution in terms of amplitude.)

Plain old telephone system (POTS)   The analog public switched telephone system we all know and love.

Plain text   Digital or analog signals that are not encrypted and from which the information can be extracted relatively easily.

Point-to-point   A communication link or transmission between only two stations.

Point-to-point link   A data communication link connecting only two stations.

Point-to-point protocol (PPP)   Protocol that allows a computer to use TCP/IP with a standard telephone line and a high-speed modem. PPP is a new standard and offers more enhanced capabilities than SLIP.

Point-to-point transmission   Transmission between two designated stations.

Primary rate   Formally known as primary rate interface (PRI). The ISDN equivalent of a T1 circuit delivers 23 B channels and a D channel or 30 B+D running at 1.544 and 2.048 Mbps, respectively. Defined for business use.

Private branch exchange (PBX)   A private telephone switching system, usually located on the user premises. Connected to a common group of lines from one or more telephone company's central offices to provide services to many users internally.

Pulse code modulation (PCM)   A process in which a signal is sampled, and each sample is quantized independently of other samples and converted by encoding to a digital signal.

Quantization   A process in which the continuous range of values of a signal is divided into nonoverlapping subranges, a discrete value being uniquely assigned to each subrange.

Quarter common intermediate format (QCIF)   A video format defined in H.261 that is characterized by 176 luminance pixels on each of 144 lines, with half as many chrominance pixels in each direction. QCIF has one quarter as many pixels as FCIF.

Raster   A predetermined pattern of scanning lines within a display space, for example, the pattern followed by an electron beam scanning the screen of a television camera or receiver.

Reliable transmission   Connection-mode data transmission which guarantees sequenced, error-free, flow-controlled transmission of messages to the receiver.

Resolution   A measurement of the smallest detail that can be distinguished by a sensor system under specific conditions. For video equipment, often measured in terms of pixels.

Resource reservation protocol (RSVP)   Provides receiver-initiated setup of resource reservations for multicast or unicast data flows, with good scaling and robustness properties.

Restricted channel   A digital communications channel for which each increment of p gives a useful capacity of only 56,000 bits per second, instead of 64,000 bits per second. This is currently common in North America and was originally due to a ones density limitation in T1 circuits.

RGB   Acronym for red-green-blue. A connection that consists of three different signals used to carry the red, green, and blue elements of a color image. Since the image is unencoded, it results in higher resolution and picture clarity than that allowed by NTSC video (which contains composite, encoded color information). Three different lines are needed for connection instead of the one for NTSC signals.

RJ-11   Acronym for registered jack number 11. It is the standard modular phone jack for the United States.

Routing table protocol (RTP)   Used in Banyan VINES routing with delay as a routing metric.

RS-232   A serial interface standard for transmission of unbalanced signals between a variety of computer, media, and multimedia peripherals. It transmits at a maximum of 19.2 Kbps and uses a 25-pin connector.

RS-422   A serial interface standard for transmission of balanced and unbalanced signals between a variety of higher end computer, media, and multimedia peripherals. It allows a maximum data rate of 10 Mbps.

RS-449   A serial interface standard for transmission of balanced and unbalanced signals between a variety of higher end computer, media, and multimedia peripherals. It allows a maximum data rate of 10 Mbps and uses a 37- or 9-pin connector.

Sampling rate   The number of samples taken per unit time; the rate at which signals are sampled for subsequent use, such as modulation, coding, quantization, or any combination of these functions.

Security   The condition achieved when designated information, material, personnel, activities, and installations are protected against espionage, sabotage, subversion, and terrorism, as well as against loss or unauthorized disclosure. The term is also applied to those measures necessary to achieve this condition and to the organizations responsible for those measures. With respect to classified matter, it is the condition that prevents unauthorized persons from having access to official information that is safeguarded in the interests of national security.

Simplex operation   Operation method in which transmission occurs in only one preassigned direction.

Subband adaptive differential pulse code modulation (SB-ADPCM)   A process that splits the audio frequency band into two subbands (higher and lower), and the signals in each subband are encoded using ADPCM.

Synchronous   A process where the information and control characters are transmitted at even intervals in order to preserve continuity (synchronization) within a data communications system.

T1 carrier   A system using time division multiplexing to carry 24 digital voice or data channels each at 64 Kbps (DS-0) over copper wire. Total speed is 1.544 Mbps which is called DS-1 or digital signal level one.

Telecommunication   Any transmission, emission, or reception of signs, signals, writings, images, and sounds or information of any nature by wire, radio, visual, or other electromagnetic systems.

Teleconferencing   The use of electronic communication to allow two or more people at two or more locations to visually and/or aurally interact with each other. Common methods of real-time teleconferencing include videoconferencing, audioconferencing, audiographic conferencing, and business television.

Teleconferencing system   A collection of equipment and integral components (customer premise equipment and facilities) required to process teleconferencing programs and control data, less network interface devices.

Teletraining   (Also known as distance learning, teleseminar, or electronic classroom.) The use of teleconferencing point-to-point or multipoint to provide interactive remote site training.

Terminal   An endpoint which provides for real-time, two-way communications with another terminal, gateway, or MCU. A terminal may provide: speech only, speech and data, speech and video, or speech, data, and video.

Terminal equipment   A device or devices connected to a network or other communications system used to receive or transmit data. It usually includes some type of I/O device.

Token ring   A ring type of local area network (LAN) in which a supervisory frame, or token, must be received by an attached terminal or workstation before that terminal or workstation can start transmitting.

Toll quality   (3-kHz analog bandwidth) ordinary telephone voice quality.

Transmission control protocol (TCP)   The TCP/IP standard transport level protocol that provides the reliable, full-duplex, stream service on which many application protocols depend. TCP allows a process on one machine to send a stream of data to a process on another. It is connection-oriented in the sense that before transmitting data, participants must establish a connection.

User A person, organization, or other entity that employs the services provided by a telecommunication system for transfer of information to others.

User datagram protocol (UDP) Unreliable networking layer which sits at same level of networking stack as TCP.

Video That portion of a signal that is related to moving images.

Video codec *See* codec.

Videoconference *See* video teleconferencing.

Videoconferencing *See* video teleconferencing.

Video teleconferencing Two-way electronic form of communications that permits two or more people in different locations to engage in face-to-face audio and visual communication. Meetings, seminars, and conferences are conducted as if all of the participants are in the same room.

Videotelephony Relating to videophones and video teleconferencing.

Visual display unit A device with a display screen, usually equipped with a keyboard, for example, a cathode ray tube display, light-emitting diode display, liquid crystal display, or plasma panel.

Wideband That property of a circuit having a bandwidth greater than 4 kHz. In the case of wideband audio, G.722 specifies a bandwidth of 7 kHz.

Windowing Capability to divide the video display into two or more separate regions with displays from different sources in each region. For example, one window could display data, another one motion video of the remote site, another one graphics, and another one motion video of the home site.

About the Author

Richard A. Schaphorst, the founder of Delta Information Systems, Inc., obtained his B.S.E.E. degree from Lehigh University and worked at Philco Ford Aerospace in the area of image communications prior to founding Delta. In recent years, he has been one of the key leaders in the development of standards for facsimile and teleconferencing. He is a member of, and contributor to, the ANSI standards committee T1A1.5. He was recently designated by the UN/ITU as the leader (Rapporteur) to develop the H.324 series of standards for videoconferencing/videophone over the public switched telephone and mobile networks.

Mr. Schaphorst was one of three Core Members of the U.S. delegation to the Specialists Group for Coding the Visual Telephony, which developed the H.261 Recommendation for the codec operating at P × 64 Kbps ISDN rates. Mr. Schaphorst is a member of the U.S. delegation to the MPEG meetings.

He has published articles and given numerous presentations on the subject of image communication and standards. Recent publications include an IEEE Press offering entitled "Teleconferencing" and a book (co-authored with Dennis Bodson) entitled *Digital Fax Technology and Applications.*

Mr. Schaphorst has served as chairman of the Facsimile Equipment and System Standards Committee (TR-29) of the EIA and has been awarded three patents in the area of image processing.

Index

Computer Telephony Integration, Second Edition, Rob Walters

Convolutional Coding: Fundamentals and Applications,
Charles Lee

Desktop Encyclopedia of the Internet, Nathan J. Muller

*Distributed Multimedia Through Broadband Communications
Services,* Daniel Minoli and Robert Keinath

Electronic Mail, Jacob Palme

*Enterprise Networking: Fractional T1 to SONET, Frame Relay to
BISDN,* Daniel Minoli

FAX: Digital Facsimile Technology and Applications, Second Edition,
Dennis Bodson, Kenneth McConnell, and Richard Schaphorst

Guide to ATM Systems and Technology, Mohammad A. Rahman

Guide to Telecommunications Transmission Systems,
Anton A. Huurdeman

A Guide to the TCP/IP Protocol Suite, Floyd Wilder

Information Superhighways Revisited: The Economics of Multimedia,
Bruce Egan

International Telecommunications Management, Bruce R. Elbert

Internet E-mail: Protocols, Standards, and Implementation,
Lawrence Hughes

Internetworking LANs: Operation, Design, and Management,
Robert Davidson and Nathan Muller

Introduction to Satellite Communication, Second Edition,
Bruce R. Elbert

Introduction to Telecommunications Network Engineering,
Tarmo Anttalainen

Introduction to Telephones and Telephone Systems, Third Edition,
A. Michael Noll

LAN, ATM, and LAN Emulation Technologies, Daniel Minoli and
Anthony Alles

The Law and Regulation of Telecommunications Carriers,
Henk Brands and Evan T. Leo

Videoconferencing and Videotelephony: Technology and Standards, Second Edition, Richard Schaphorst

Visual Telephony, Edward A. Daly and Kathleen J. Hansell

World-Class Telecommunications Service Development, Ellen P. Ward

For further information on these and other Artech House titles, including previously considered out-of-print books now available through our In-Print-Forever® (IPF®) program, contact:

Artech House	Artech House
685 Canton Street	46 Gillingham Street
Norwood, MA 02062	London SW1V 1AH UK
Phone: 781-769-9750	Phone: +44 (0)171-973-8077
Fax: 781-769-6334	Fax: +44 (0)171-630-0166
e-mail: artech@artech-house.com	e-mail: artech-uk@artech-house.com

Find us on the World Wide Web at:
www.artechhouse.com